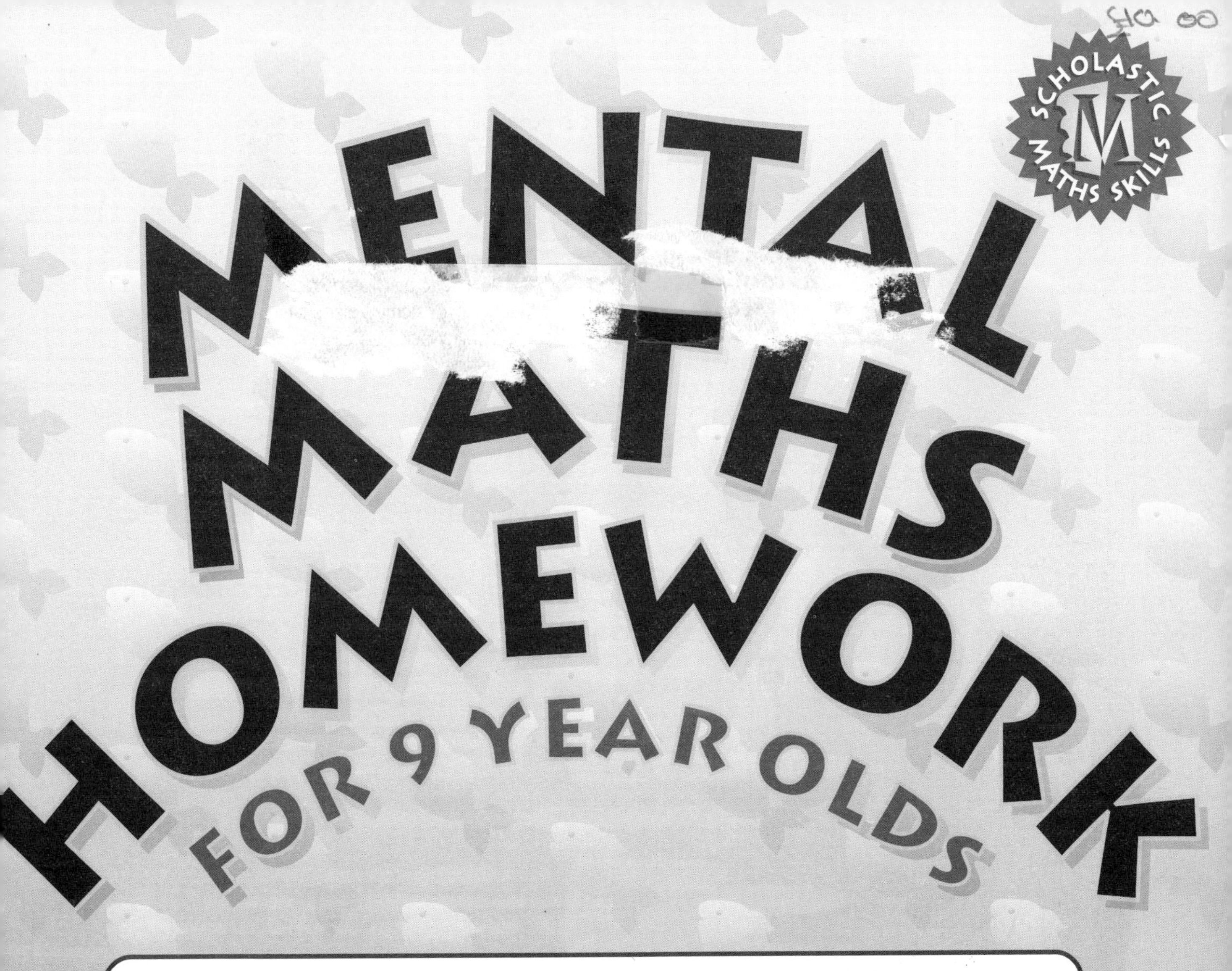

SERIES EDITOR
Lin Taylor
The IMPACT Project, University of North London

AUTHOR
Jill Meyer

EDITOR
Joel Lane

ASSISTANT EDITOR
Clare Miller

SERIES DESIGNER
Anna Oliwa

DESIGNER
Paul Cheshire

ILLUSTRATIONS
Garry Davies

COVER ARTWORK
James Alexander/David Oliver
Berkeley Studios

Designed using Adobe Pagemaker
Published by Scholastic Ltd, Villiers House, Clarendon Avenue, Leamington Spa, Warwickshire CV32 5PR

1 2 3 4 5 6 7 8 9 0   9 0 1 2 3 4 5 6 7 8

British Library Cataloguing-in-Publication Data
A catalogue record for this book is available from the British Library.

ISBN 0-439-01705-X

CONTENTS

IMPACT

ABOUT HOMEWORK

Homework can be a very useful opportunity to practise and develop children's understanding of the work done in school. Games and maths challenges can be very good activities to share with someone at home, especially to develop mental maths strategies and maths language skills. Research* indicates that parental involvement is a major factor in children's educational success. Most parents want to help their children with their school work, but often do not know how and 'traditional' homework does not involve parents. Shared homework activities, such as can be found in *Mental Maths Homework*, are designed to be completed with a parent or helper, such as a sibling, neighbour or other adult who can work with the child. Working one-to-one with an adult in the home environment really has a powerful effect. The National Numeracy Strategy strongly supports this type of homework, which is in line with a variety of government guidelines on the role of parents and making home links.

ABOUT MENTAL MATHS AT HOME

Mental Maths Homework is particularly concerned to develop children's *mental* mathematics. In order to become competent at mental calculation, children need to talk about mathematics and try out different strategies, as well as to practise number facts and skills. Children explaining their mathematics to a parent or helper can help to clarify and develop their understanding of the mathematics. This type of homework, developed by The IMPACT Project, is a *joint* activity: the helper and child working together.

ABOUT MENTAL MATHS HOMEWORK

This series comprises of six books, one for each age group from 6–11 years (Year 1/P2–Year 6/P7). Each book contains 36 photocopiable activities – enough for one to be sent home each week throughout the school year, if you wish. The activities concentrate on the number system and developing children's calculation strategies and are designed to fit into your planning, whatever scheme you are using. Since these books are designed to support the same aims of developing mental maths strategies and vocabulary, they make an ideal follow-on to the class work outlined in Scholastic's other *Mental Maths* series. The objectives for each activity are based on those in the National Numeracy Strategy *Framework for Teaching Mathematics* and the content is appropriate for teachers following other UK curriculum documents.

USING THE ACTIVITIES IN SCHOOL

Although the books are designed for a particular age group they should be used flexibly so that the right level of activity is set for a child or class. All the activities are photocopiable: most are one page, some are two, or require an extra resource page (to be found at the back of the book) for certain games or number card activities. The activities for older children will generally take longer than those for younger children.

BEFORE

It is essential that each activity is introduced to the class before it is sent home with them. This fulfils several crucial functions. It enables the child to explain the activity to the parent or carer; ensuring the child understands the task. It also familiarises the child with the activity; developing motivation and making the activity more accessible. This initial introduction to the activity can be done as part of a regular maths lesson, at the end of the day, or whenever fits in with your class's routine.

AFTER

It is also important that the child brings something back to school from the activity at home. This will not necessarily be substantial, or even anything written, since the activities aim to develop mental mathematics. It is equally important that what the child brings in from home is genuinely valued by you. It is unlikely that parents will be encouraged to share activities with their children if they do not feel that their role is valued either. Each activity indicates what should be brought back to school, and the teachers' notes (on pages 5–8) offer guidance on introducing and working with or reviewing the outcome of each activity.

HELPERS

All the activities have a note to the helper explaining the purpose of the activity and how to help the child, often emphasizing useful vocabulary. The helpers' notes also give indications of how to adapt the activity at home, and what to do if the child gets stuck. Many of the activities are games or fun activities which it is hoped that the parent and child will enjoy doing together and will do again, even when not set for homework, thus increasing the educational benefit. It is particularly beneficial for a game to be played a number of times.

OTHER WAYS TO USE THE ACTIVITIES

The activities offered in *Mental Maths Homework* are very flexible and will be used in different ways in different schools. As well as being used for shared homework, they could form the basis of a display or a school challenge, or be used as activities for a maths club. Or, they could be used independently of the school situation by parents who wish to provide stimulating and appropriate educational activities for their children.

USING THE ACTIVITIES AT HOME

If you are a parent using these activities outside of school:

- Choose an activity you both think looks interesting and get going straight away with your child. Make the work *joint*: the helper and the child working out what has to be done *together*.
- Read the instructions to your child and ask him or her to explain what has to be done. It is very effective for the child to do the explaining.

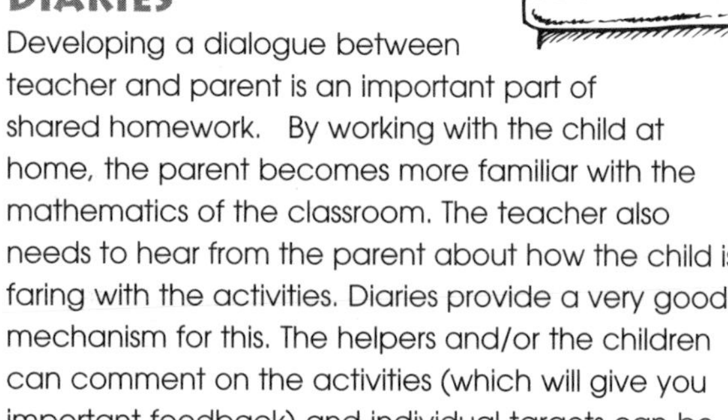

USING HOMEWORK DIARIES

Developing a dialogue between teacher and parent is an important part of shared homework. By working with the child at home, the parent becomes more familiar with the mathematics of the classroom. The teacher also needs to hear from the parent about how the child is faring with the activities. Diaries provide a very good mechanism for this. The helpers and/or the children can comment on the activities (which will give you important feedback) and individual targets can be put into the diary. The diaries can act, therefore, as an important channel of communication. (See below for details about finding out more information about diaries.)

ABOUT THIS BOOK

Mental Maths Homework for 9 year olds focuses on using and applying the knowledge that the children have been practicing in mental maths. The content emphasis is on numbers to 10 000 and on using the number system by looking for sequences and patterns. The children will also use number in calculations, employing the four operations (addition, subtraction, multiplication and division) as they see fit. Finally, they will need to make sense of problems by putting the above skills together in a coherent and useful way; this book will help them continue to do that through games and fun challenges.

Although the main emphasis is on mental maths, some activities ask the children to use pencil and paper procedures – not as given routines, but through their understanding of the problem and the calculations they see as necessary to extend their thinking. Much attention is still given to discussion and explanation of how they reach results and conclusions, which is why the partner is still an essential part of these activities.

---

* Bastiani, J. & Wolfendale, S. (1996) *Home-School Work: Review, Reflection and Development* David Fulton Publishers.

THE IMPACT PROJECT

The activities in *Mental Maths Homework* have all been devised by members of The IMPACT Project, based at the University of North London. The project, a pioneer of shared homework, with a wealth of experience, is involved in a variety of initiatives concerning parental involvement and homework. It also supports schools in setting up a school framework for shared homework. If you would like help with developing shared homework, planning a whole-school framework for homework or developing mental mathematics at home and at school, maybe through INSET with experienced providers, contact The IMPACT Project. Information about other activities undertaken by the project and about other IMPACT books and resources, such as the IMPACT diaries, is also available from The IMPACT Project.

The IMPACT Project
University of North London
School of Education
166–220 Holloway Road
London
N7 8DB

tel. no. 020 7753 7052

fax. no. 020 7753 5420

e-mail: impact-enquiries@unl.ac.uk
impact-orders@unl.ac.uk

web: http://www.unl.ac.uk/impact

COUNTING & ORDERING

ROUNDING UP AND DOWN

OBJECTIVE: To round whole numbers to the nearest 10 or 100.

BEFORE: Use this activity to assess the children's understanding of place value. Ask them to write numbers that have, for example, 230 as their nearest ten. They need to practise this, and to understand the difference between the **nearest** and the **next** ten.

AFTER: Use Velcro numbers to make an interactive display for groups to use as an ongoing activity. As the children become more confident, larger numbers could be used.

IS IT ODD? IS IT EVEN?

OBJECTIVE: To recognize odd and even numbers to 1000 and some of their properties, including the outcomes of the sum and difference of pairs of odd and even numbers.

BEFORE: The children need to know what odd and even numbers are – and, more importantly, which digits tell you whether the number is even or odd.

AFTER: Check that the children know to look only at the units digit to see whether a number is odd or even. Can they recognize the properties of odd and even numbers? Can they express their ideas clearly in words? Encourage them to try subtracting as well.

SECRET NUMBERS

OBJECTIVES: To read and write whole numbers to 10 000, knowing what each digit represents. To partition a number into thousands, hundreds, tens and ones.

BEFORE: This activity is linked to 'Fractions in a line' (see page 13). Play the game as a class (probably split into two ability groups, so that appropriate numbers may be used). Do the children give correct answers? You may need to ask them to write down the number, so that you can give support where necessary. The 'Bet you can't' challenge is a useful activity to practise in a few spare minutes!

AFTER: Record the children's numbers on a class number line. This could lead to work on the uses of fractions and decimals.

NEGATIVE NUMBER NECKLACE

OBJECTIVE: To recognize negative numbers.

BEFORE: Use a number line to demonstrate negative numbers. A lot of counting forwards and backwards is essential for this.

AFTER: Can the children extend a number line to find numbers above and below 0? Keep a number line on display in the classroom. Apply the concept of negative numbers to temperature: practise reading a thermometer that is showing negative values.

FRACTIONS IN A LINE

OBJECTIVE: To order simple fractions.

BEFORE: This activity is linked to 'Secret numbers' (page 11). Lots of class carpet maths using a number line, ordering numbers on the line and reading aloud forwards and backwards, will help the children to grasp the idea.

AFTER: Can the children order the fractions? They need to have lots of practice with equivalent fractions. Can they find a common denominator and recognize denominators in a multiplication pattern? Remember that the children need to be able to read and write fractions in both numerical and word form.

FOLDING FRACTIONS

OBJECTIVE: To recognize the equivalence of simple fractions.

BEFORE: A class lesson on folding paper is useful preparation for this. The children need practice in seeing parts of a shape and identifying the fractions that the shape is divided into. They might carry out this activity on their own; or small groups might try making different folds with the same paper shape.

AFTER: Encourage the children to look for patterns in the equivalent fractions. Can they make a fraction that is bigger or smaller by multiplying and dividing (as in $\frac{1}{4} \times 3 = \frac{3}{4}$)?

CREATE IT

OBJECTIVES: To use decimal notation. To know what each digit in a decimal represents. To order a set of decimals.

BEFORE: The children will need lots of practice with place value to tackle this activity. It could be linked with 'Easy rule 10' (page 26) when decimals are being discussed. Do the children remember to put in the decimal point? Money is a useful teaching aid here. Again, counting aloud with number lines is essential.

AFTER: This activity needs to be repeated often, with the aid of a visual number line. Start with the mixed fractions the children made, as these are the numbers they feel most comfortable with.

ADDITION & SUBTRACTION

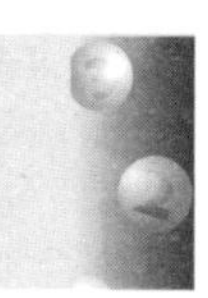

BINGO 10S

OBJECTIVE: To add or subtract 1, 10, 100 or 1000 to/from any numbers.

BEFORE: To play the game with a 1 to 6 dice and add on tens, you need to use numbers from 21–126. To play the game with a 1 to 6 dice and add on hundreds, you need to use numbers from 111–666. Learning about the layout of a Bingo card will also aid children's understanding of tens numbers. Throughout the year, practise counting on and back in tens from any number with the aid of a number grid.

AFTER: This is a popular class game; after initial practice, the children can play it without adult help. Point out that some numbers cannot be made, and so the game can never be won.

RACE TO 500

OBJECTIVE: To use knowledge of number facts and place value to add (or subtract) a pair of numbers mentally.
BEFORE: The children will need a lot of practice in adding (or subtracting) numbers in their heads. Model the activity a few times with the class.
AFTER: Try some of the children's games; make sure they explain them to each other. Ask them to write their rules down for others to use, then allow time for them to play their games. If they haven't designed games, let them try to do so in small groups.

BACKWARD SENTENCES

OBJECTIVE: To use the relationship between addition and subtraction.
BEFORE: Model the game, especially the building of number sentences. This needs a lot of practice.
AFTER: Work on some number sentences that are more complex than those used in the game. Ask the children to draw snake-shaped number sentences (without answers) and share them with the class. Who can find the answers? Display some new number sentences each day for the children to answer.

FUN SHAPES

OBJECTIVES: To use the relationship between addition and subtraction. To use the vocabulary of addition.
BEFORE: Demonstrate with some shapes the process that the children will need to work through.
AFTER: Working with the class or small groups, display the children's shapes with one corner number missing. Can they find the missing numbers? Can they write down the calculation(s) they had to do?

FOOD SHOPPING

OBJECTIVE: To use knowledge of number facts and place value to add or subtract a pair of numbers mentally.
BEFORE: Model the activity with your own shopping list. Talk through the two stages: adding up five prices, then taking the total away from £10.00. Discuss the vocabulary of 'change'.
AFTER: Use the shopping lists for fun addition. Emphasize the idea of looking for known number bonds to make 10 or 100. An interactive display of questions could be made.

FUNCTION ADDITION

OBJECTIVES: To develop rapid recall of addition and subtraction facts. To use the relationship between the two operations. To develop mental strategies for adding (or subtracting) 9 or 99 by adding (or subtracting) 10 or 100 and adjusting by one.
BEFORE: Function machines are a good way to demonstrate the effect of performing an operation on a number. Demonstrate how a function machine works. Make sure the children understand the task.
AFTER: Discuss ways of adding or subtracting 9, 19, 29 and so on. Talk about strategies. Can the children see any patterns? Test the children's function machines: take input numbers from the class and ask the child whose machine is being tested to give outputs. Who can work out the function?

RINGING AROUND

OBJECTIVE: To develop rapid recall of addition and subtraction facts.
BEFORE: Work through the activity using the school phone number. Encourage the children to look for known number facts to help them.
AFTER: Try out some of the children's number sentences as mental calculations. Did any of the children have the same target number; if so, did they make different sentences? Again, use the children's number sentences to make 'sentence snakes'.

ROCKET LANDING

OBJECTIVES: To use the relationship between addition and subtraction. To use the vocabulary of subtraction.
BEFORE: Give the children a number to keep in their heads, then ask them to take another number away from it. Encourage them to take away the tens, then count back the units. Demonstrate this with a 1–100 grid: show a spider crawling vertically to take away tens and horizontally to take away units. The children need to understand that the counter goes in the space with the correct tens digit, and that they have to remember the exact number.
AFTER: This could be a class game that is played regularly – perhaps with higher numbers, or by subtracting or multiplying the dice numbers.

WORD SUMS

OBJECTIVE: To use the mental addition strategies of putting the largest number first and identifying doubles.
BEFORE: Discuss the mental addition strategies described on the sheet. Practise them separately and as part of rehearsing the homework. Ask the children to say how they are using the strategies.
AFTER: Who has the highest-value word? Working with small groups, put the children's words in order of numerical value; then join the groups together to encourage co-operative working. If the alphabet values are reversed, so that Z = 1 and A = 26, can the children work out whose word has the highest value?

SUBTRACTION PYRAMID

OBJECTIVE: To use knowledge of number facts and place value to add or subtract a pair of numbers mentally.

BEFORE: Demonstrate the activity with a different set of numbers, so that the children get the idea. This will revise number bonds.
AFTER: Discuss the effect of putting 1000 in the top box. This links to 'Easy rule 10' (page 26). What patterns did the children find?

MULTIPLICATION & DIVISION

EASY RULE 10

OBJECTIVE: To multiply or divide whole numbers less than 1000 by 10 and understand the effect.
BEFORE: Practice in multiplying by 10 is essential. Use pictures, cubes and other visual aids to demonstrate the change when a number is multiplied by 10. Encourage the children to look for patterns.
AFTER: This activity helps children to express the rule for multiplying by 10 (not 'adding a 0' but 'moving all the numbers one place to the left'). Can they multiply and divide any decimal number by 10?

MULTIPLICATION BINGO

OBJECTIVE: To know multiplication facts by heart and derive the corresponding division facts quickly.
BEFORE: Practise chanting the 5× table forwards and backwards. Use fingers to count the number of 5s. Try playing the game.
AFTER: Play this game with other times tables. The game can also be reversed: the children write down the numbers 1–9 and the cards have the answers.

CORNER VALUE

OBJECTIVE: To know multiplication facts by heart and derive the corresponding division facts quickly.
BEFORE: Make sure the children understand that all the corners of a shape in the activity must have the same number. Try a few examples.
AFTER: Which numbers did the children find more than one way to work out? Can they write an operation sentence such as: $3 \times 5 = 15$?

DICE AND TABLES

OBJECTIVE: To know multiplication facts by heart and derive the corresponding division facts quickly.
BEFORE: Play the game on the carpet with the class; set the numbers on the dice as appropriate.
AFTER: Can the children work out all the possible combinations from the two dice? Again, change the dice to make a particular pattern emerge or to make the game easier or harder.

MONEY CORNERS

OBJECTIVES: To use the relationships between subtraction and addition and between multiplication and division. To use the inverse operation to check answers.
BEFORE: Using different numbers in a variety of shapes, model the activity. Demonstrate that more than one answer is possible. Remind the children that they can only use the real coins that are in circulation.
AFTER: Share some of the ways of making the totals. Change some of the totals by amounts such as 10p. Ask: *What effect will that have on the answer?*

TABLE ANSWERS

OBJECTIVE: To know the multiplication facts from the 2, 3, 4, 5 and 10 times tables.
BEFORE: Practice chanting the tables. Use fingers to represent the multiplier within a table: three fingers displayed when talking about the 5-times table means 3×5. Work through the activity, making sure the children understand the timing aspect of the game.
AFTER: Were the children able to use more than one multiplication table to find more than one way of making some numbers? The game could be used to practise any times table, or to recall division facts.

TIME FOR MY NAME

OBJECTIVES: To use the relationships between addition and multiplication and between division and multiplication. To use the vocabulary of time.
BEFORE: Discuss the number of seconds in a minute, half a minute, two minutes and so on, until the children get the idea of multiplying and dividing with time. Try the activity with a few names; make sure the children can explain what they are doing.
AFTER: Extend the activity by asking: *How many times could you write your name in an hour?*

TIME SHARE

OBJECTIVES: To understand division and its relationship to subtraction and multiplication. To use the vocabulary of division. To know when to round up or down after division.
BEFORE: The children will need to be familiar with remainders. Mental division work needs to be practised with an understanding that division is the inverse of multiplication. So to find out what $50 \div 2$ is, you can also say: *How many 2s in 50?*
AFTER: Discuss how you can predict which numbers will divide equally. Try the same activity with other total numbers. Encourage the children to check their answers using multiplication.

WHAT'S THE PATTERN?

OBJECTIVE: To know doubles and halves, or derive them rapidly.
BEFORE: Much practice will be needed for the children to understand halving. They should know that halving and doubling are inverse operations, and should know the basic doubling facts up to 2×10.
AFTER: Discuss with the children what pattern they have noticed. Can they work out the next number? Can they do the odd numbers? What have they found out about halving an odd number? Why is this?

MULTISTEP & MIXED OPERATIONS

DOUBLE EDGE

OBJECTIVE: To use related number facts and doubling or halving in mental calculation.
BEFORE: Practise adding sets of numbers to 100. Practise doubling. Model the activity for the children.
AFTER: Ask individual children to come and draw the first face of their shape (with the numbers) on the board. Can other children work out the other visible faces? A 2D shape pattern is useful for this.

ADDING THE TABLES

OBJECTIVE: To solve a mathematical problem using mental calculation strategies and jottings.
BEFORE: The children need to be very familiar with all the times tables. Recite them together, using fingers (see 'Table answers' above). Make the number grids by copying and adapting the grid on page 47; or ask the children to make them.
AFTER: Discuss which numbers can be made in more than one way. Ask the children to write down all the combinations that make those numbers. Encourage them to check with each other what numbers they couldn't make. Discuss these numbers. The activity could continue in class over a long period of time.

CHANGES

OBJECTIVE: To use a variety of operations and strategies for mental calculation.
BEFORE: This activity is similar to 'What's your rule?' (page 46), but the rules used are simpler and are provided as cards. Work through some examples.
AFTER: Discuss which change cards were hard to work out, and any new cards the children made.

HIT THE SPIDER

OBJECTIVE: To use a variety of operations and strategies for mental calculation.
BEFORE: Practise making some simple target numbers in different ways. Encourage the children to think of all the different operations they could use.
AFTER: Compare the ways the children have found to make a target number. Devise new ways. This could be a ten-minute activity done at various times.

1 TO 10

OBJECTIVES: To know, and recall rapidly, addition and subtraction facts. To know multiplication facts and derive the corresponding division facts quickly.
BEFORE: Encourage the children to try with their own three numbers (to 10).
AFTER: Look at the numbers that can be made in more than one way. Discuss which combination of numbers gave the most numbers from 1–10. The game could also be played with four or five numbers to 20.

MAIL ORDER CATALOGUE

OBJECTIVES: To solve number problems. To recognize and explain patterns and relationships.
BEFORE: Provide some catalogues; encourage the children to add the prices of several items. Make sure that the children understand what 'half price' and 'change' mean. Practise halving and doubling.
AFTER: You could prepare a table with priced items and questions such as *How many items could you buy with £10.00?* Discuss the strategies the children used to work out their change from the half price items.

TARGET PRACTICE

OBJECTIVES: To add several one-, two- and three-digit numbers together in a problem-solving context. To explain their methods and reasoning.
BEFORE: The children need to have practice in adding money, to know which coins are in use, and to have experience of doubling and multiplying by 3.
AFTER: Look at how the children recorded their ways of making £5.00. How did they know when they had found all the possible ways?

THE FACTS OF NUMBER

OBJECTIVES: To use different operations flexibly and in quick succession to solve a problem.
BEFORE: The children need practice with different number operations. Make sure they understand the range of choices they can make in the activity.
AFTER: Compare the children's 100 squares. Which numbers have been missed out? This activity could be continued in class over a long period of time.

UNDER THE COVERS

OBJECTIVE: To choose and use appropriate operations to solve numerical problems.
BEFORE: The children need experience of playing this game (perhaps just using multiplication) and practice in finding and writing number sentences to make a given number. Both of these activities can be covered within the ten-minute daily mental maths sessions.
AFTER: The children can share their grids and covers. This game can be used to practise any operations, and is a good activity for an independent group.

WHAT'S YOUR RULE?

OBJECTIVE: To choose and use appropriate operations to solve numerical problems.
BEFORE: The children must be confident in giving 'start' and 'finish' numbers after carrying out an operation. This activity links with 'Function addition' (page 21) and 'Changes' (page 38). If the children find making their own rules hard, demonstrate by writing the number sentences.
AFTER: Find each other's rules as a class or in groups. Which rules are harder to work out? Why is this?

NAME **DATE**

ROUNDING UP AND DOWN

YOU WILL NEED: a helper, a pencil.

YOU ARE GOING TO: round numbers to the nearest ten and hundred.

❑ Look at the numbers given on the sheep. Can you write the nearest ten and the nearest hundred for each number?

❑ Take turns with your helper to write each answer, then check each other's work at the end.

	Nearest 10	Nearest 100
286		
691		
314		
729		
532		
153		
465		
847		
978		

❑ Bring your completed sheet back to school to be checked.

BET YOU CAN'T
Make up some numbers and ask your helper to find the nearest ten and the nearest 100 to them.

DEAR HELPER

THE POINT OF THIS ACTIVITY: is to help your child get a feel for **approximation** (rounding numbers). Then, when working with number problems, he or she can make a reasonable guess at the answer and so judge whether his or her actual answer is likely to be right.

YOU MIGHT LIKE TO: ask your child to write as many numbers as he or she can which have (for example) 430 as their nearest ten. He or she will need to know the five numbers just above 430 and the five numbers just below 430. Try using numbers above 1000: if your child has a good understanding of how the number system works, he or she should still be able to tell you the numbers.

IF YOU GET STUCK: start with numbers below 100, as this will build your child's confidence. If necessary, draw a 1–100 number square or number line, so that your child can see which 'ten' is the nearest.

Please sign: . . . .

NAME DATE

IS IT ODD? IS IT EVEN?

YOU WILL NEED: a helper, paper and pencils.

YOU ARE GOING TO: investigate the effect of adding and subtracting odd and even numbers.

❑ Write down ten 3-digit numbers on pieces of paper.

❑ Now sort them into odd and even numbers.

❑ Add two even numbers together: ☐ + ☐ = ☐

Add two odd numbers together: ☐ + ☐ = ☐

Add an even number and an odd number: ☐ + ☐ = ☐

❑ What do you notice about these answers? Think about which answers are odd and which are even. Explain to your helper which answers are odd and which are even, and why you think this is.

❑ Together, check your idea with other pairs of numbers.

❑ Bring your ideas about the answers back to school for discussion.

YOU MIGHT LIKE TO TRY

❑ Subtracting one odd number from another: ☐ – ☐ = ☐

❑ Subtracting one even number from another: ☐ – ☐ = ☐

❑ Subtracting an odd number from an even number: ☐ – ☐ = ☐

❑ Subtracting an even number from an odd number: ☐ – ☐ = ☐

What do you notice about these answers?

DEAR HELPER

THE POINT OF THIS ACTIVITY: is to help your child to make general statements about odd and even numbers and to give examples to prove their points. Encourage your child to talk about how he or she knows whether a number is odd or even, and to say what else he or she knows about odd and even numbers. For example:

- The last digit of an even number is 0, 2, 4, 6 or 8.
- After 1, every second number is odd.
- The numbers on both sides of an odd number are even.
- If you add two odd numbers together, you get an even number.
- The difference between two odd numbers or two even numbers is even.

YOU MIGHT LIKE TO: encourage your child to try multiplying with small numbers, using a calculator. Ask your child to say what happens when you multiply two odd or two even numbers. What happens if you multiply an odd number by an even number?

IF YOU GET STUCK: work with two-digit numbers and use a calculator to do the additions. You are not checking your child's ability to add up, but his or her ability to recognize odd and even numbers and see patterns in the addition of these numbers. To do this, your child needs to understand the properties of odd and even numbers.

Please sign: . . . .

NAME DATE

SECRET NUMBERS

YOU WILL NEED: a helper, paper and pencils.

YOU ARE GOING TO: try to guess the secret number on the number line.

❑ Think of a number between 1 and 10 000. Write it down on a piece of paper.

❑ Tell your helper when your 'secret number' is ready. He or she has to try to guess your number, but you can only answer with the word **higher** or **lower**.

For example, with the secret number 2500.

Your helper says:	You answer:
1000	▲ higher ▲
2000	▲ higher ▲
3000	▼ lower ▼
2850	▼ lower ▼
2425	▲ higher ▲
2500	▶ correct ◀

❑ Now it is your helper's turn to write down a secret number.

❑ Bring to school some numbers that you, or your helper, took a lot of guesses to find.

BET YOU CAN'T

Play 'Find the Digit'. When you have tried the game above several times, put all the 'secret numbers' face up in the middle of the floor. To win each number, you have to answer a question such as: *Which number has 3 units? Which number has 6 tens? Which number has 4 hundreds?*

DEAR HELPER

THE POINT OF THIS ACTIVITY: is to help your child appreciate the size of a number in relation to other numbers. This will help him or her to make better estimates of the answers when working out number problems. It will also improve his or her understanding of what each **digit** in a number represents (hundreds, tens, units and so on).

YOU MIGHT LIKE TO: try fractions and decimal numbers, if you feel your child is ready for this. Offer some clues, perhaps reminding your child of guesses that he or she has already made: 'I've said it's more than 510 but less than 511.'

IF YOU GET STUCK:

- Write down all the guesses on a number line, so that your child can start to track down the 'secret number' visually.

1000... > 2000... > 2425.... > (2500) < ...2850 < ...3000

- To begin with, stick to numbers less than 1000.

Please sign: . . . .

NAME DATE

NEGATIVE NUMBER NECKLACE

YOU WILL NEED: a helper, paper and pencils.

YOU ARE GOING TO: order some positive and negative numbers.

❑ With your helper, look at the pattern of the number stones on this necklace.

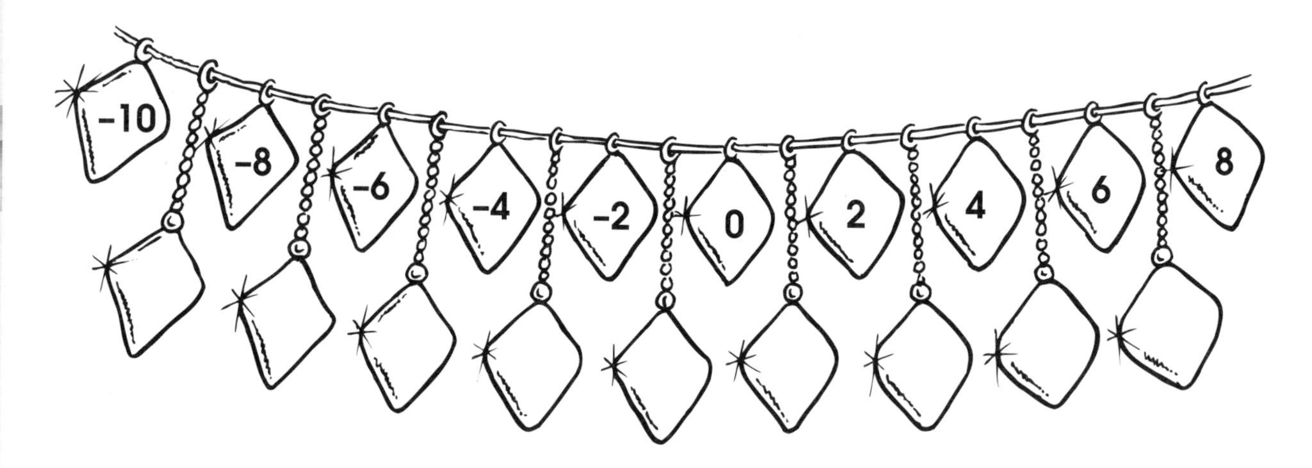

❑ Now look at the missing stones drawn below. Take turns to copy one into the correct place on the necklace.

❑ Bring back the finished necklace to school.

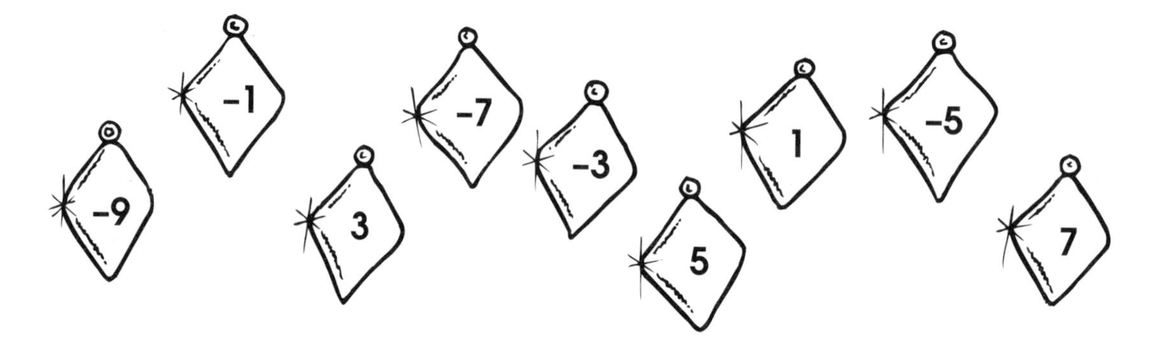

BET YOU CAN'T

Make up some stones to place in a necklace of your own.

DEAR HELPER

THE POINT OF THIS ACTIVITY: is to help your child to understand, read and write negative numbers. He or she will use negative numbers when talking about temperatures and money, and will need to be able to find the difference between two such numbers in order to work out changes.

YOU MIGHT LIKE TO: draw some number lines which include negative numbers, say from –8 to 8, and then ask your child to say at which point on each of the lines he or she would place a number such as –4.

–8 8

IF YOU GET STUCK: a visual number line will be very useful to help your child understand the idea of negative numbers.

Please sign: . . . .

NAME DATE

FRACTIONS IN A LINE

YOU WILL NEED: a helper, a pencil and small pieces of paper, a piece of string.

YOU ARE GOING TO: make a fraction number line.

❑ Can you make a number line using the fractions written here? Copy each fraction out on to a separate piece of paper, then put them in order.

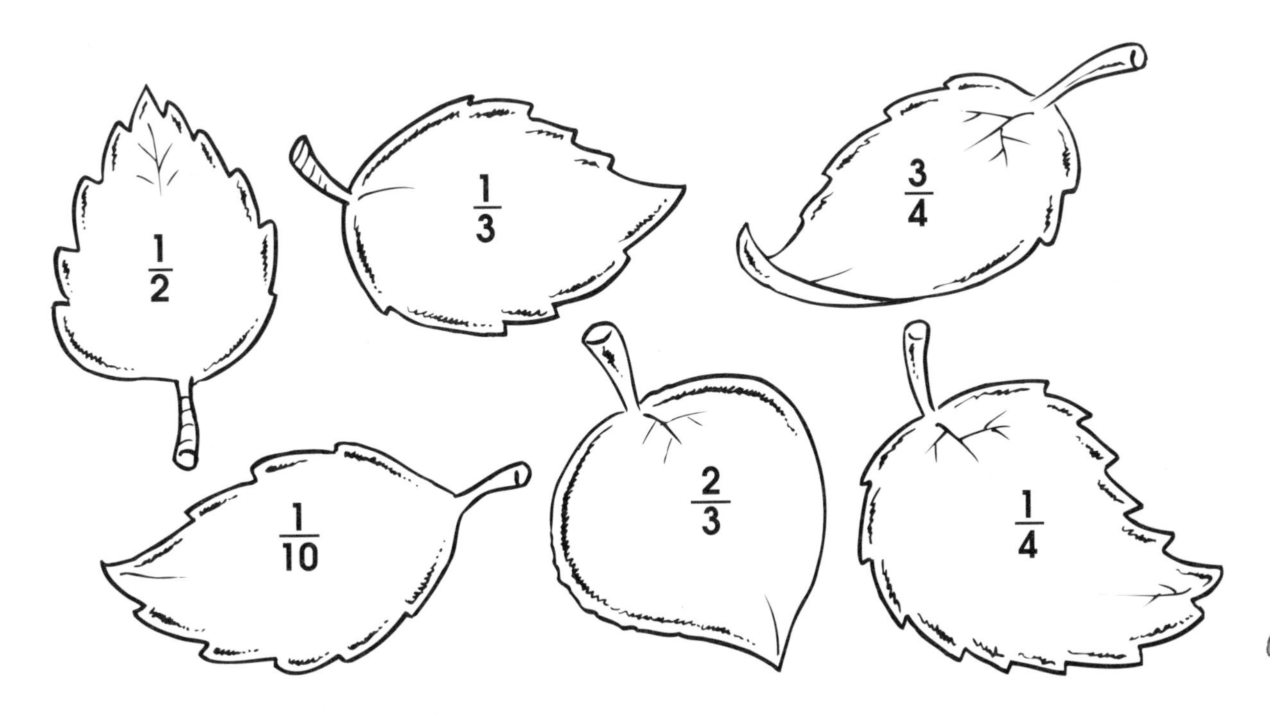

❑ When you have made your number line, thread a piece of string through it to keep the numbers in order. Bring it back to school.

HANDY HINT!

You need to think: 'What value is this fraction?' A good place to start is whether the fraction is greater or less than a half.

BET YOU CAN'T

Make up some more fractions of your own and challenge your helper to order them on a number line.

DEAR HELPER

THE POINT OF THIS ACTIVITY: is to help your child to put fractions in order. This will extend his or her understanding of the sizes of different numbers. By Year 4, your child should be able to read and write numbers, including fractions, in both numerical and word form.

YOU MIGHT LIKE TO: make up some fractions and encourage your child to order them.

IF YOU GET STUCK: encourage your child to identify those fractions greater than, equal to or less than ½. If your child draws out the lower number of the fraction as dots and puts a circle around the upper number of the fraction, he or she should be able to compare different fractions visually.

Please sign: . . . .

NAME DATE

FOLDING FRACTIONS

YOU WILL NEED: a helper, a rectangular strip of paper, a pencil.

YOU ARE GOING TO: compare fractions to see which ones have the same value.

❑ Fold a strip of paper in half and shade in half of it, like this:

❑ Fold it again. What fraction does the shaded part of the paper show now? ($\frac{2}{4}$ or two quarters.)

❑ Now fold it again. How many eighths are in the shaded half?

❑ Can you see a pattern in these numbers? Talk to your helper about it.

❑ Can you work out how many 256ths would equal ½? You will need to do this using the number pattern, not folding the strip of paper.

❑ Bring your answer into school, with an explanation of how you worked it out.

BET YOU CAN'T

Keep doubling the number until you are in the thousandths.

DEAR HELPER

THE POINT OF THIS ACTIVITY: is for your child to recognize fractions which have the same value. He or she will use this knowledge later to put fractions in order and recognize simple relationships between them. Ask your child to write down the fractions in his or her explanation using both numbers and words, so that he or she can get used to both forms.

YOU MIGHT LIKE TO: repeat the activity, working out fractions that are the same as a quarter (¼). From this, your child could tell you what fractions are the same as ¾ by multiplying the ¼ number by 3.

IF YOU GET STUCK: use several paper strips of the same size and shape, so that your child can physically place half of a strip on top of the original shaded half each time and see how many quarters, eighths and so on fit the space.

Please sign: . . . .

CREATE IT

YOU WILL NEED: a helper, scissors, a pencil, paper, some Blu-Tack, the number cards 0–9 (cut from this sheet).

YOU ARE GOING TO: make and order decimal numbers.

- ❑ Cut out the number cards on the edge of this sheet.
- ❑ Using these cards, and drawing in a decimal point, make a decimal number equivalent to each of the **mixed fractions** below.
- ❑ Stick the decimal numbers next to the fractions, using Blu-Tack. Don't forget to put in the decimal point!

0
1
2
3
4
5
6
7
8
9

$3\frac{6}{10} =$

$5\frac{9}{10} =$

$2\frac{4}{10} =$

$1\frac{7}{10} =$

$7\frac{10}{10} =$

HANDY HINT! You will only need to use each number card once.

- ❑ Discuss with your helper what the value of each digit in the decimal numbers is.
- ❑ Write the five decimal numbers in order of size, from smallest to largest.
- ❑ Bring the ordered decimals back into school to display as a number line.

BET YOU CAN'T

Use the number cards 0–9 again to make up some other mixed fractions. Remember: you can only use each number card once.

DEAR HELPER

THE POINT OF THIS ACTIVITY: is to help your child to understand, read and write decimals. Children need to know the significance of **decimal places** – for example, that the value of 6 in 3.6 is six tenths. This will help them to put decimal numbers in order and to appreciate their size.

Discuss the idea of **decimal places** with your child, and remind him or her about the decimal point.

YOU MIGHT LIKE TO: extend this activity by asking your child to convert pounds into pence and vice versa. For example:

- *Write 364p in £.*
- *How many pence in £6.74? How many in £8.25?*
- *Write in pounds the total of ten £1 coins, six 10p coins and three 1p coins.*

IF YOU GET STUCK: complete the task together with the aid of a number line. This will help your child to understand what 'tenths' are.

1 1.1 1.2 1.3 1.4 1.5 1.6 1.7 1.8 1.9 2

Please sign: . . . .

NAME DATE

BINGO 10S

YOU WILL NEED: a helper, perhaps another person (as 'caller'), pencils, paper or thin card, scissors, three dice, the blank Bingo card on this sheet.

YOU ARE GOING TO: add tens and hundreds to numbers.

❑ Make enough blank Bingo cards (like the one shown below) for the people who want to play.

❑ Fill in the cards. For each card, you and your helper need to think of ten numbers between 21 and 126 to write in the squares. You can shade in any five squares.

❑ The caller needs to have three dice. He or she rolls two dice and says the number: 'A five and a three, 53.' He or she then rolls the third dice and says the number on it. The players must add this many **tens** to the original number to find the 'Bingo' number (so if the third dice number is 4, 40 must be added to 53, giving a total of 93).

❑ If you have this total number on your board, you can cross it out. The caller must keep a record of the numbers thrown and the 'Bingo' numbers made.

❑ When somebody wins by crossing off all the numbers in a line, the caller has to check that the numbers that are crossed off are, in fact, correct.

❑ If you like, go on to see who can make a 'Full house'.

❑ Make some Bingo cards to bring back to school, so that you can play this game in class.

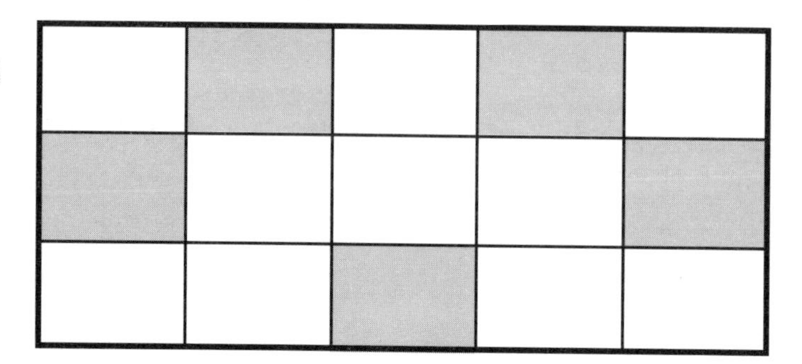

BET YOU CAN'T

Play this game using numbers between one hundred and eleven (111) and six hundred and sixty-six (666) on your Bingo card. When you roll the third dice, add on that many **hundreds**.

DEAR HELPER

THE POINT OF THIS ACTIVITY: is to help your child to have a greater understanding of number. He or she will need to read and write the numbers, know what each digit represents and work with tens (or hundreds) to play the game. If necessary, you or your child can be the caller as well as playing the game.

YOU MIGHT LIKE TO:

- Make a few Bingo cards in advance.
- Make a 10 × 11 number grid with the numbers 21–126 written on, to help the caller record the number totals.
- Encourage your child to put the numbers on the cards in column groups, as on a real Bingo card (so that a particular number, if it appears, will always be in the same column). The columns would represent twenties, starting from 21 and ending at 126: 21–41, 42–62, 63–83, 84–104, 105–126. (Remember that you cannot make numbers with a units digit of 7, 8, 9 or 0.) This will give further practice in ordering numbers.

IF YOU GET STUCK: practise rolling the dice together and adding on the tens. Use a number square to demonstrate what happens when you add on 10.

Please sign: . . . .

NAME DATE

RACE TO 500

YOU WILL NEED: a helper, a pack of cards, pencils and paper.

YOU ARE GOING TO: add up numbers to reach 500.

❑ Shuffle the cards and place them face down in a pile. Take the top two cards and add them together. (All picture cards count as 10.) Then your helper can do the same.

❑ Work out whose total is higher. That player goes first the next time.

❑ Keep on taking two cards, adding them together and adding this number to your last total. Remember that the player with the higher overall total goes first each time.

❑ The first player to reach a score over 500 is the winner. You and your helper should keep a running total on paper.

❑ Take your running totals back to school.

BET YOU CAN'T

❑ Start at 500 and add two cards together as before, but take the total away from 500 and bring your total down until you reach 0.

❑ Design a version of this game in which the two cards are multiplied together. Bring your game back to school to share with the class.

DEAR HELPER

THE POINT OF THIS ACTIVITY: is to help your child practise mental strategies for addition (and subtraction). This involves skills such as: looking for number bonds; adding the tens and then the units; rounding up or down and then adjusting by the appropriate number. Encourage your child to use appropriate vocabulary such as **more, add, plus, sum, total** and **altogether.**

Don't allow your child to write down the whole sum. He or she should be using the paper for recording, not for working out.

Reversing the game by starting with 500 and subtracting is harder, but similar skills should be used: taking away tens and then units; rounding numbers and adjusting. The vocabulary is important as well: **subtract, take away, how many are left, how much less, difference between.**

YOU MIGHT LIKE TO: play the game to 1000 for more practice.

IF YOU GET STUCK:

- Write down each sum, so that your child doesn't have to remember it.
- Encourage your child to add on the tens and then the units. A 1–100 square could be useful for showing this: 64 + 27 = 64 + 20 = 84, 84 + 7 = 91.

Please sign: . . . .

NAME **DATE**

BACKWARD SENTENCES

YOU WILL NEED: a helper, pencils and paper, a calculator.

YOU ARE GOING TO: make number sentences for known answers.

❑ To start, you and your helper both write down any number from 1 to 9.
❑ Show each other your numbers and make the **largest** two-digit number you can from them. For example, if you wrote down 6 and your helper wrote down 3, the number you made would be 63.

Now the race is on.

❑ Make up a number sentence with that number as the answer – for example, 40 + 23 = 63. It could be addition or subtraction.
❑ Whoever is first to complete a number sentence reads it out, and the other player checks it on a calculator.
❑ If the sentence is correct, the player gets a point. The winner is the first player to collect 10 points.
❑ Bring your number sentences back to school.

BET YOU CAN'T

Work out a number sentence that has addition **and** subtraction in it.

DEAR HELPER

THE POINT OF THIS ACTIVITY: is to encourage your child to use his or her knowledge that addition and subtraction are **inverses** of one another. For example, if 25 + 15 = 40 then 40 – 25 = 15 and 40 – 15 = 25. This activity requires your child to work backwards: start with the answer, then subtract a number from it, then use the result to construct a correct number sentence.

YOU MIGHT LIKE TO: set a rule that the number sentence must have three or more parts to it. This will make your child do more mental addition and subtraction each time.

IF YOU GET STUCK: take away the competitive element and work through a few number sentences with your child, demonstrating how he or she can check them with the calculator.

Please sign: . . . .

FUN SHAPES

YOU WILL NEED: a helper, a pencil and paper, the shapes on this page.

YOU ARE GOING TO: make some addition shapes.

- ❑ Choose a shape from those shown on the right and draw it. Now choose a number between 1 and 1000 to write in the middle of your shape.

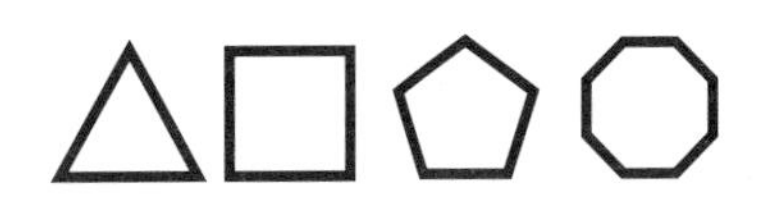

- ❑ Take turns to write a number on each corner so that the shape's corners add up to make the number in the middle. For example:

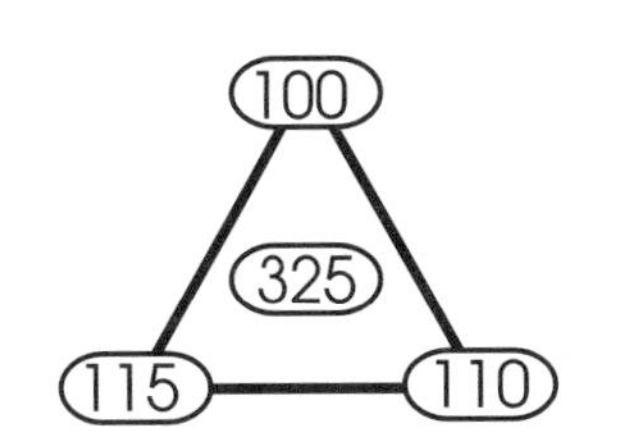

- ❑ Now try the other shapes. Talk about these questions:
Does it matter which corner you start with?
Does it matter which way round the shape you go?
- ❑ Bring the shapes you have drawn to school, with the numbers written on the corners and inside the shapes.

YOU MIGHT LIKE TO TRY
Drawing some shape pictures or irregular shapes with numbers inside for your helper to try. Compare his or her solutions with yours.

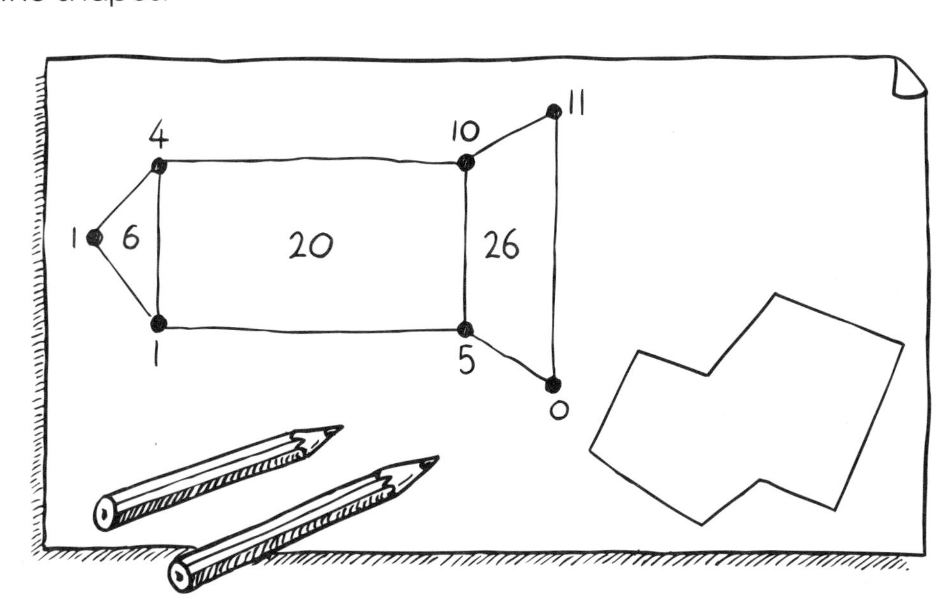

DEAR HELPER

THE POINT OF THIS ACTIVITY: is to help your child to see that numbers can be added together in any order without affecting the result. This means that 74 + 36 = 36 + 74; and that 34 + 12 + 16 = (34 + 16) + 12 = 34 + (12 + 16). Your child also needs to understand that the **inverse** of addition is subtraction: if 46 + 12 = 58, then 58 - 12 = 46 and 58 - 46 = 12. To find solutions to the problems above, your child will need to add numbers together and/or subtract them from his or her target number.

Use a range of vocabulary when talking to your child - for example: *How many more do you need?... What is the sum total of those numbers?*

YOU MIGHT LIKE TO: encourage your child to use numbers up to 10,000 and to avoid using numbers which have 0 in the units place. This will make it harder to add and subtract the numbers, giving your child more practice in these skills.

IF YOU GET STUCK: start with target numbers up to 100 and make them divisible by 10 or 5. Then, as your child's confidence grows, increase the number size and the difficulty of the units digit. Encourage him or her to add the tens first and then add on the ones.

Please sign: . . . .

NAME DATE

FOOD SHOPPING

YOU WILL NEED: a helper, a shop or supermarket receipt (from family food shopping), a pencil and paper.

YOU ARE GOING TO: work out what you can buy with your money.

❑ Look at the shop receipt together and choose your five favourite items (priced under £2 each). Write down their prices.

❑ Add these prices together. How much change would you be given if you gave the shopkeeper £10.00?

❑ Do this again with five different items.

❑ Which five items on the shop receipt would give you the most change from £10? How much change would you get?

❑ Suppose you only had £10 to spend. Which five items on the list would give you the least change from £10? How much change would you get?

❑ Now think about how you worked out the amounts of change. Write down how you calculated the change and bring this note back to school, along with the shopping receipt.

BET YOU CAN'T

Add up the total cost for the whole list. Ask your helper to tear off the bottom part of the receipt first!

HANDY HINT!

Look for things which can be added together to make a pound. Add all the 1ps first, then the 10ps, then the pounds.

DEAR HELPER

THE POINT OF THIS ACTIVITY: is for your child to find strategies to work out addition and subtraction problems quickly. Children find money a little more difficult than 'pure' numbers, even though they are working with the same number system.

This activity also encourages your child to explain how he or she worked out the answers. This will make your child think harder about what he or she is actually doing. Find out what strategies are being used by discussing the process as your child carries out the additions and subtractions. Help your child to write down his or her ways of working.

YOU MIGHT LIKE TO: Use a catalogue to work out together what your child might spend £100 on.

IF YOU GET STUCK:

- Remind your child that 100 pence is the same as a pound.
- Help him or her to look for lower- and higher-priced items respectively when answering the questions above.
- Let your child work with two or three low-priced items to begin with.
- Encourage the strategy suggested in the 'Handy hint' above. Discuss what it is doing to reinforce the language.

Please sign: . . . .

FUNCTION ADDITION

YOU WILL NEED: a helper, a pencil, some paper, some blank cards, a shoe box.

YOU ARE GOING TO: use rules such as 'add 9' to change numbers. On its first setting, this function machine always adds 9. So when the input is 20, the output is 29.

❑ Work out the outputs for 36, 51, 47, 88 and 127. Fill them in on the table below.

❑ On its second setting, the machine always subtracts 9. So when the input is 20, the output is 11. Work out the other outputs with the second setting. Write them in the table.

❑ On its third setting, the machine always subtracts 19. So when the input is 20, the output is 1. Work out the other outputs with the third setting and complete the table.

❑ Look down the columns. Can you see a pattern in your answers?

Input	Add 9	Subtract 9	Subtract 19
20	29	11	1
36			
51			
47			
88			
127			

❑ Invent a function machine yourself. Draw up another table and ask your helper to fill in the columns.

❑ Now use an old shoe box to make your function machine. Add some cards to put in and pick out. Bring your machine back to school, but don't say what the function is. Let the class work it out from your input and output numbers.

BET YOU CAN'T

Work out an easy way to add or subtract 9, 19, 29, 39 and so on.

DEAR HELPER

THE POINT OF THIS ACTIVITY: is to help your child to use quick mental strategies for solving near-10 addition and subtraction problems. This will allow him or her to calculate more quickly and gain a feeling for sensible answers. Your child also needs to check each answer to an addition or subtraction with the other operation. For example, 179 – 48 can be worked out as 179 – 50 + 2 = 131. This can be checked by working out 131 + 48 = 131 + 50 – 2 = 179.

YOU MIGHT LIKE TO: give your child some other numbers to put into the function machine and time his or her answers. You could play too and see who can do more in one minute.

IF YOU GET STUCK: use numbers less than 100, and start by just adding or subtracting 10.

Please sign: . . . .

NAME DATE

RINGING AROUND

YOU WILL NEED: a helper, a pencil, some paper, a timer, your telephone number.

YOU ARE GOING TO: use the total of the digits in your phone number to make number sentences.

- ❑ Write down the last part of your telephone number (without the area code). It may have six, seven or eight digits.
- ❑ Add up the total of the digits. So if your number is 887799, the total is 8 + 8 + 7 + 7 + 9 + 9 = 48.
- ❑ Now you and your helper have to write down as many six-figure (or seven- or eight-figure) numbers as you can in one minute whose digits add up to give **the same total** (in this case, 48).
- ❑ Can you find a quick way to work out the numbers? Discuss what methods you are using.

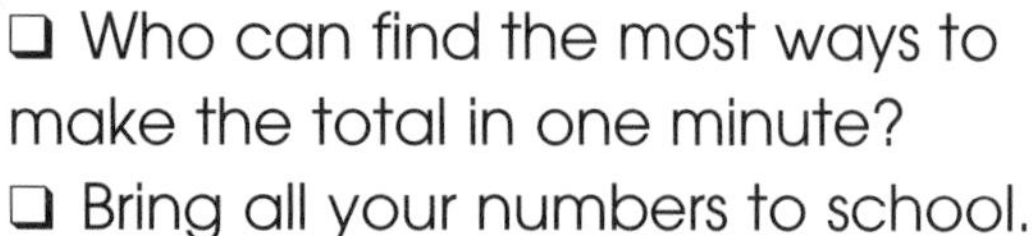

- ❑ Who can find the most ways to make the total in one minute?

- ❑ Bring all your numbers to school.

BET YOU CAN'T

Find a seven-figure number by using seven subtractions from the starting number 75 to reach your phone number total. For example: 75 – 3 – 2 – 3 – 8 – 9 – 2 = 48, so the number 323892 is an answer for 48.

DEAR HELPER

THE POINT OF THIS ACTIVITY: is to help your child consolidate his or her knowledge of addition and subtraction facts by working out different ways to make a total. He or she will need to think about **number bonds**: 2 + 8 = 10, so 12 + 8 = 20 and 20 – 8 = 12. While adding up digits, your child will also have to take the total away from the original target. For example, if your child has four digits that add up to 24 and the target is 30, he or she needs to take 24 away from 30 to see that the last two digits need to add up to 6.

YOU MIGHT LIKE TO: encourage your child to use subtraction, rather than addition, to find the answers.

IF YOU GET STUCK:

- Suggest that your child writes down the additions and keeps a running total as he or she goes along.

  8 + 6 + 2 + 4 + 9 + 7

  14 16 20 29 36

- Encourage your child to look for patterns, such as pairs of numbers which add up to 10 or 15.

Please sign: . . . .

NAME DATE

ROCKET LANDING

YOU WILL NEED: a helper, two counters, two dice, pencils, paper, the rocket on this sheet.

YOU ARE GOING TO: throw dice to make sums in a game.

- ❑ Both you and your helper put your counters on 100 at the top of the rocket.
- ❑ Take turns to throw two dice and add the two numbers together.
- ❑ Take the total away from 100. **Write down the number you get.** So if you threw 4 and 3 (a total of 7), your new number is 100 – 7 = 93.
- ❑ How many tens does your number have? Move your counter into the correct space for that many tens. So if your new number is 93, you move your counter into the space marked 90.
- ❑ Keep on doing this, moving your counter down when you reach the next ten after subtracting a dice total.
- ❑ To finish, you must throw the exact number on the dice that you need to reach 0.

- ❑ Don't forget to write down all the numbers you make as you play.
- ❑ Bring these numbers back to school. Be prepared to explain how you got them and which squares you landed on.

100
90
80
70
60
50
40
30
20
10
0

BET YOU CAN'T

- ❑ Find the difference between the two numbers on the dice and take it away from 100.
- ❑ Make a rocket from 1000 down to 0. You could multiply the numbers on the two dice so that the game doesn't take too long.

DEAR HELPER

THE POINT OF THIS ACTIVITY: is to help your child with his or her understanding of subtraction. He or she will need to add two numbers, then take the total away from a larger number and notice when the next ten has been reached.

Help your child to understand and use different vocabulary for subtraction, such as **take away, subtract, how much less, how many are left, difference between** and **minus**. Remind your child of the subtractions he or she has done, using a variety of terms.

YOU MIGHT LIKE TO: encourage your child to write down the subtractions he or she does, as this will improve his or her recording skills.

IF YOU GET STUCK: try drawing a larger rocket with all the numbers 0–100 on it. Your child can then move the counter back space by space to consolidate the idea that subtraction causes a number to become smaller.

Please sign: . . . .

NAME DATE

WORD SUMS

YOU WILL NEED: a helper, a pencil and paper.

YOU ARE GOING TO: add numbers by looking for pairs that make 10 and doubles that you know.

❑ How much is a word worth? **A = 1, B = 2, C = 3 and so on up to Z = 26.**
❑ Can you work out the values of some words?

For example, the word FRIEND is worth 6 + 18 + 9 + 5 + 14 + 4.
To work out this total, rearrange the numbers, starting with the largest first. So the sum now becomes 18 + 14 + 9 + 6 + 5 + 4.
Now look for number bonds and doubles:
18 + (14 + (9 + 5)) + (6 + 4)
= 18 + 28 + 10 = 28 + (18 + 10)
= 28 + 28 = 56

❑ Take turns to give each other words and work out how much they are worth. Show on paper how you worked them out.
❑ Bring all your words into school, along with your working out of the word totals.

HANDY HINT!
Remember to start with the highest number first, then look for useful doubles and number bonds.

BET YOU CAN'T

❑ Find a really long word with z in it. This will make it difficult for your helper!
❑ See what is the highest-value word you can find.

DEAR HELPER

THE AIM OF THIS ACTIVITY: is to encourage your child to use two particular mental strategies for addition. To be a good mathematician, your child needs to be ready to use a range of mental strategies. The strategies encouraged here are: put the largest number first, then look for number bonds (pairs of numbers that make a useful total such as 10) and doubles.

YOU MIGHT LIKE TO: give your child some difficult words to find the 'value' of, such as **wristwatch** or **velociraptor**.

IF YOU GET STUCK:
- Go through each word value problem with your child to make sure that he or she is using the strategies suggested.
- If this still proves too difficult, just give your child the sums and let him or her practise ordering the numbers. Include some numbers that make doubles, to encourage him or her to look more carefully. Don't worry too much about the value of the word: concentrate on the maths skills.

Please sign: . . . .

SUBTRACTION PYRAMID

YOU WILL NEED: a helper, a pencil, the pyramid on this sheet.

YOU ARE GOING TO: make a pyramid of numbers by adding and subtracting.

In this pyramid, **the two numbers above each brick add to make the number inside that brick.**

- ❑ Can you work out the numbers that are missing?
- ❑ With your helper, take turns to write a missing number until the pyramid is complete.
- ❑ Talk to your helper about the patterns you can see in the brick. Look across and diagonally.
- ❑ Be ready to talk in class about the patterns you saw.

100						
100	100					
100	200	100				
100			100			
100				100		
100					100	
100	600	1500	2000	1500	600	100

BET YOU CAN'T

- ❑ Continue the pattern with another line of squares along the bottom of the pyramid.
- ❑ Predict what numbers you will get along the bottom if you put 1000 in the top square.

DEAR HELPER

THE POINT OF THIS ACTIVITY: is to help your child improve his or her quick recall of number bonds when working in tens and crossing 1000. This will help your child to estimate when solving problems – for example, to evaluate how 'sensible' an answer is.

There are three different strategies for solving this problem: addition, subtraction and looking for patterns. Your child may use any one (or more than one) of these methods.

YOU MIGHT LIKE TO:

- Ask your child to explain how he or she found the answers. What patterns can he or she see? Can he or she continue the pyramid?
- Provide some addition and subtraction puzzles where the number added to or taken away from is a multiple of 10. For example: 120 + 19 = ? 240 + 58 = ? 170 – 77 = ?
- Alternatively, you could ask 'How many to the next 100?' For example: 360 + ? = 400 or 585 + ? = 600.

IF YOU GET STUCK:

- Write in a few more numbers to demonstrate the pattern. Talk about how you can work out the answer in a variety of ways (see above).
- Ask your child to make a pyramid with 10 at the top.

Please sign: . . . .

NAME **DATE**

EASY RULE 10

YOU WILL NEED: a helper, a 1–100 number grid (see page 47), red and blue coloured pencils.

YOU ARE GOING TO: multiply and divide by 10 in a game.

❑ Decide with your helper which of you is the blue team and which is the red team.

❑ Secretly choose a number from the 1–100 number grid and multiply it by 10.

❑ Tell your helper your grid number multiplied by 10. So if the number you chose was 30, you would tell your helper '300'.

❑ Can your helper say what number you started with? If so, he or she can colour in that number square on the grid with their colour.

❑ Now your helper chooses a number **not coloured in** and tells you the number multiplied by 10. If you can guess the number, you get to colour in that square with your colour.

❑ When all the squares are coloured in, stop. The winner is the player whose colour covers more squares.

❑ Is there a rule for multiplying by 10? Bring your ideas back to school.

BET YOU CAN'T

❑ Play the game on a 101–200 number grid.

❑ Multiply by 100 instead of 10.

DEAR HELPER

THE POINT OF THIS ACTIVITY: is to give your child the opportunity to practise multiplying and dividing by 10. To give you a problem to work out, he or she first has to multiply by 10; but to work out what your number is, he or she needs to divide by 10.

Encourage your child to establish a rule for multiplying by 10. Suggest that he or she thinks about the numbers **moving one place to the left**, rather than adding a 0 to the end, because the latter idea can be misleading when children start to multiply decimals.

YOU MIGHT LIKE TO: try with decimals – but make sure your child knows the rule for multiplying by 10 first.

IF YOU GET STUCK: start with numbers to 10 only, and write down both the original number and its 'multiply by ten' number so that your child can see the pattern over several goes.

Please sign: . . . .

MULTIPLICATION BINGO

YOU WILL NEED: a helper, a pencil and paper, a set of 0–9 'Digit cards' (see page 47), a copy of the 5 times table.

YOU ARE GOING TO: play Bingo by multiplying numbers by 5.

❑ First, you and your helper both write down any five numbers from the 5× table (up to 10 × 5). For example, you might write down 5, 20, 35, 40 and 50. Don't write down the same numbers as your helper.

❑ Now shuffle the digit cards and take turns to turn over a card. Multiply the number you turn over by 5. (The 0 card counts as 10.) Call out the result.

❑ If you have written that number as one of your Bingo numbers, you can cross it out.

❑ Keep going until you have crossed all your numbers out and can shout 'Bingo!'

❑ Check the 'Bingo!' call with the cards that have gone and the five numbers that you wrote out. If no mistakes have been made, you are the winner!

❑ When you have finished, write down the tables you have used like this: 1 × 5 = 5 2 × 5 = 10, and so on.

Bring your tables back to school.

BET YOU CAN'T

❑ Play with another multiplication table, such as the 2, 3, 4 or 10 times table.

❑ Play the game the other way round by writing down five numbers from 1 to 10 and then turning over cards that have multiplication answers on them (5, 10, 15 and so on).

DEAR HELPER

THE POINT OF THIS ACTIVITY: is to help your child to learn multiplication facts and use them quickly. Your child needs to understand the question *How many 5s in 30?* as well as *5 times what is 30?*

YOU MIGHT LIKE TO:

- Time this activity. When you turn over each card, give your child a time limit within which he or she has to work out the answer. This will improve your child's quick mental recall of multiplication facts.
- Play the game with the corresponding division facts. This will help your child to use division skills when solving numerical problems.

IF YOU GET STUCK:

- Encourage your child to write out the table first, then gradually remove this support.
- Concentrate on the first five multiplication facts in a table.
- Limit the tables to 2×, 5× and 10×.

Please sign: . . . .

NAME DATE

CORNER VALUE

YOU WILL NEED: a helper, a pencil and the 'Target number shapes' on page 29.

YOU ARE GOING TO: find the answer to some division shape puzzles.

❑ Look at the two shapes below. Each shape has a target number in it. The first shape has the same number on each corner. When the number of corners is multiplied by the number on each corner, the answer is the target number. The numbers on the second shape must follow the same rules.

❑ Can you find the corner number for the second shape?

❑ Can you find the corner numbers for all the shapes on page 29?

❑ Bring your shapes and numbers back to school for everyone to see.

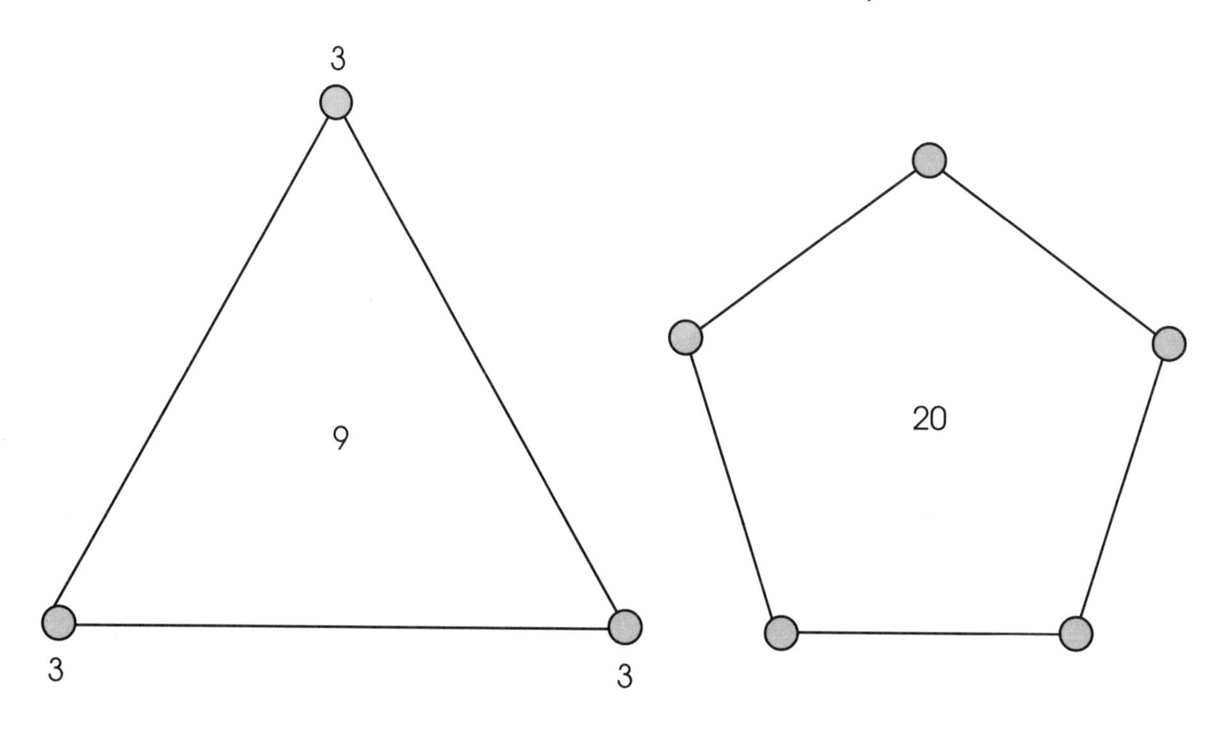

BET YOU CAN'T

❑ Work out some different answers for the same target number, but with a different shape.

❑ Make up some shapes and target numbers of your own. Make sure you know what the answers are! Bring these shapes back to school for your friends to try.

DEAR HELPER

THE POINT OF THIS ACTIVITY: is for your child to become more familiar with times tables and to recognize different multiplication sums which make the same target. One simple way to find the corner number for each shape is to divide the target number by the number of corners; this will lead your child to see the relationship between multiplication and division. When discussing the problems, use vocabulary such as **times, multiply, multiplied by, share, divide, divided by** and **divided into**.

YOU MIGHT LIKE TO: test your child by drawing some shapes with 6, 7, 8 and 9 sides, using numbers from the 6, 7, 8 and 9 times tables as the target numbers.

IF YOU GET STUCK: stick to the 2, 3, 4, 5 and 10 times tables. Ask your child to write down these tables to help them work out the problems.

Please sign: . . . .

NAME DATE

TARGET NUMBER SHAPES

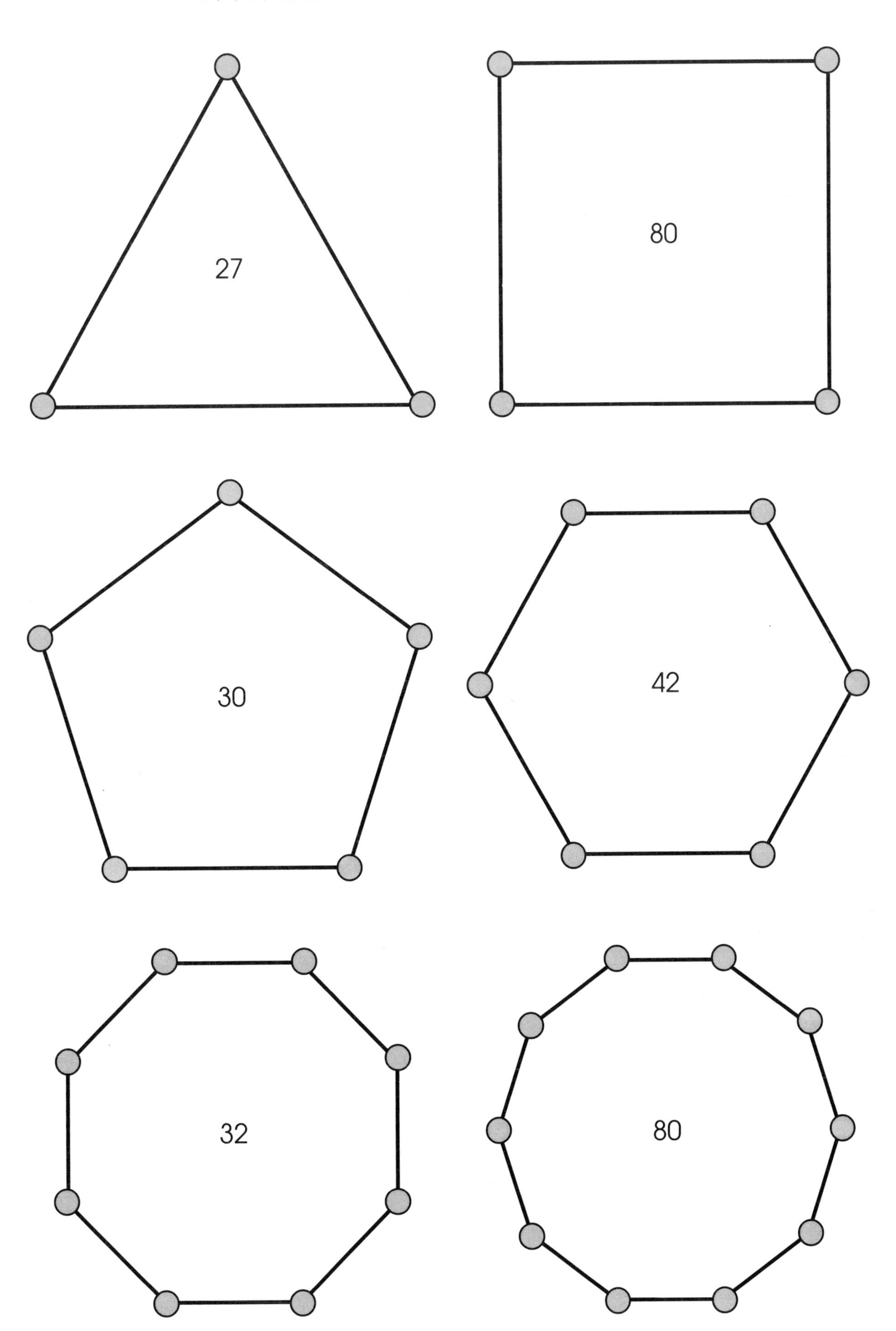

NAME DATE

DICE AND TABLES

YOU WILL NEED: a helper, a pair of dice, a pencil and paper, a stopwatch, a calculator.

YOU ARE GOING TO: see how fast you can multiply two numbers together.

❑ Be ready with your pencil and paper before you start this activity. It is all about speed!

❑ You roll one dice and your helper rolls the other. Multiply the two numbers together. Write down the number sentence and the answer as quickly as you can, while your helper times you.

❑ If you find and write down the correct answer in between 5 and 10 seconds, you score 1 point. If you get the answer in less than 5 seconds, you score 2 points. You can check your answers on a calculator if you want to.

❑ See how quickly you can score 20 points.

❑ Bring the number sentences you have written down to school.

BET YOU CAN'T

❑ Change one of the dice to read 5, 6, 7, 8, 9 and 10. This will give you slightly harder multiplications to work out. See whether you can get each answer in twenty seconds.

❑ Work out all the possible combinations that the dice could show, and find the answers.

DEAR HELPER

THE POINT OF THIS ACTIVITY: is to help your child to learn times tables in the context of a fun game. Children need to know the multiplication facts up to 10 × 10 as soon as possible, because these factors are used in much of the maths works they do. By playing this dice game, your child is working with numbers in a way that he or she already feels confident with, and will find it easy to turn the multiplication sums around (for example, the dice numbers 3 and 4 could be 3 × 4 or 4 × 3).

YOU MIGHT LIKE TO: stop the clock after each multiplication is done and recap, using language such as 'So how many 5s are in 25?' or 'So 5 times 6 is 30.'

IF YOU GET STUCK:

- Do not worry too much about the speed. You could encourage your child to find each answer in a minute, then gradually reduce the time.
- Work through the possible number combinations on the dice together before playing the game.

Please sign: . . . .

NAME DATE

MONEY CORNERS

YOU WILL NEED: a helper, a pencil and paper.

YOU ARE GOING TO: solve a money problem using your number skills and knowledge of coins.

❑ Look at the shapes below. For each shape, you need to place some coins at each corner. There are two rules:

1. The amount of money at each corner must be the same.
2. The corners must add up to make the amount inside the shape.

❑ What coins would you place at each corner of these shapes? An example has been done to help you. Discuss this problem with your helper.

❑ Can you find more than one way of making each corner amount?

❑ Bring your completed money shapes into school, so that the class can look at all the different ways of making the same totals.

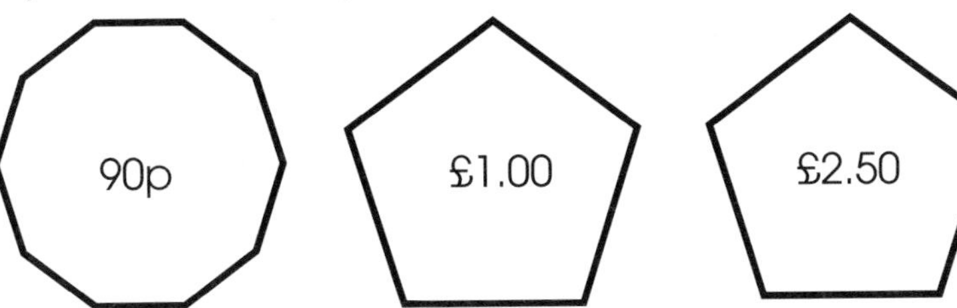

YOU MIGHT LIKE TO TRY
Making up some money shape problems of your own. Can your helper do them?

DEAR HELPER

THE POINT OF THIS ACTIVITY: is to encourage your child to use all four number operations – addition, subtraction, multiplication and division – in mental calculation strategies. To work out the answers, your child will need to divide the total amount of money in each shape by the number of corners and then break up the corner amount into coin values. To check the answers, he or she will need to add the coins on a corner and multiply that by the number of corners.

Once your child gets the idea, increase the values and use corner numbers that will need more than one coin. Don't forget to make your child check his or her answers.

YOU MIGHT LIKE TO:

- Encourage your child to check his or her answers, using inverse operations (for example, using division to check multiplication).
- Set your child some more problems of this kind, using larger amounts of money.

IF YOU GET STUCK:

- Make the numbers divisible into single coin values, such as 50p in a five-sided shape or 60p in a three-sided shape.
- Have a set of coins available for your child to look at.

Please sign: . . . .

NAME DATE

TABLE ANSWERS

YOU WILL NEED: a helper, some small pieces of scrap paper, pencils, a timer, a calculator.

YOU ARE GOING TO: play a game to test your knowledge of times tables.

- ❑ Start by writing down all the numbers in the 2 times table, starting with 0, on small pieces of paper (0, 2, 4 and so on to 24).
- ❑ Shuffle the pieces of paper. Take the top piece and write a multiplication sentence for it. For example, if the number is 10, you would write $5 \times 2 = 10$. Check your number sentence on the calculator.
- ❑ Ask your helper to time how long it takes you to write your number sentence. Start the clock when you turn the paper over. If you can write the number sentence in less than 10 seconds, and it is correct when you check with a calculator, you score a point.
- ❑ Carry on until you have scored 10 points.
- ❑ Take your pieces of paper into school to try in a maths lesson.

BET YOU CAN'T

Make this harder by using numbers from the 3, 4, 5 or 10 times table.

DEAR HELPER

THE POINT OF THIS ACTIVITY: is to help your child to recall multiplication facts in the 2 times table (and then the 3, 4, 5 and 10 times tables), including multiplication by 0 and 1. This will enable him or her to respond rapidly to oral or written questions. Start with smaller numbers (up to 10×); as your child becomes more confident and begins to see the patterns in the tables, he or she will use numbers beyond this.

As the multiplication facts become known to him or her, your child will be able to derive the corresponding division facts: '5 × 2 makes 10, so 10 divided by 2 is 5.' Help your child to build up a range of vocabulary such as **times, multiplied by** and **...lots of.**

YOU MIGHT LIKE TO: see whether your child can think of more than one answer for each number, using any times table. So 10 could be 5×2, 2×5, 1×10 or 10×1. You could award extra points for multiple correct answers.

IF YOU GET STUCK: don't worry about timing the activity. If your child has difficulty thinking up numbers, ask him or her to write out the appropriate times table for reference.

Please sign: . . . .

TIME FOR MY NAME

YOU WILL NEED: a helper, a pencil and paper, a timer.

YOU ARE GOING TO: time yourself writing your name.

❑ How many seconds are there in a minute? Tell your helper.

❑ Ask your helper to time you for 1 minute while you write your full name as many times as possible.

❑ Now work these out:
How many seconds did it take you to write your name once?
How many seconds did it take you to write each letter in your name? (Assume that you wrote at the same speed all through the minute, and that each letter took the same amount of time.)

❑ Show your answers to your helper and explain how you worked them out.

❑ Bring your answers back to school. Be ready to explain again how you found them.

BET YOU CAN'T

❑ Work out the times for writing other names in your family.

❑ Work out how many times you could write your name in 2 minutes, 3 minutes, 4 minutes or 5 minutes. If you can work out the answer for 5 minutes, how could you work out the answer for 10 minutes?

❑ Work out how many times you could write your name in 1½ minutes.

DEAR HELPER

THE POINT OF THIS ACTIVITY: is to help your child use his or her knowledge of the relationships between the four number operations (addition, subtraction, multiplication and division). The most important part of the task is for your child to explain how he or she found the answers. What strategies did he or she use?

YOU MIGHT LIKE TO: encourage your child to work out how many times he or she could write his or her name in ½ minute or 1½ minutes. This will force your child to use division in order to find the ½ minute value.

IF YOU GET STUCK: you may need to reinforce the idea (assumed in the activity) that your child is writing at a constant pace. If your child struggles with this idea, then model writing his or her name and say: '10 times in 1 minute, so that must be 20 times in 2 minutes.'

Please sign: . . . .

NAME DATE

TIME SHARE

YOU WILL NEED: a helper, an analogue clock (with numbers on), a pencil and paper.

YOU ARE GOING TO: use clock numbers to work out division problems.

- ❑ Write out all the clock's numbers, from 1 to 12.
- ❑ Both you and your helper now add up these numbers:
  1 + 2 + 3 + 4 + 5 + 6 + 7 + 8 + 9 + 10 + 11 + 12 = ?
- ❑ Did you both get the same total?
- ❑ Predict which of the numbers 1 to 12 will **divide equally** into the total.
- ❑ Now try to divide each number into the total. Were your predictions correct?
- ❑ List those numbers that have a remainder. Now write all your division answers to the **nearest** whole number. Remember that if the remainder is half or more of the dividing number, the answer is rounded **up**.

- ❑ Could you predict which numbers are more likely to divide equally into any large number? Talk this idea over with your helper, and be ready to share your thoughts with the class.
- ❑ Bring your answers back to school.

BET YOU CAN'T

Check your answers by multiplying and then adding on the remainder.

DEAR HELPER

THE POINT OF THIS ACTIVITY: is to help your child to understand the operation of division either as **sharing equally** or as **repeated subtraction**, and to understand the concept of a **remainder**.

It is important to use vocabulary such as **share, group, divided by, divided into** and **is divisible by** with your child, so that he or she has the right words to describe how he or she has done a piece of work. Your child will need to understand about remainders when rounding numbers in order to give approximate answers.

YOU MIGHT LIKE TO: Go through the activity again, using a 24-hour clock.

IF YOU GET STUCK: Provide 78 small objects (such as buttons, dried peas or counters) for your child to group. This will enable him or her to see the remainders physically.

Please sign: . . . .

WHAT'S THE PATTERN?

YOU WILL NEED: a helper, a pencil, scrap paper.

YOU ARE GOING TO: practise halving numbers.

❑ Think of a number which has a units digit of 2 (such as 42). If you halve it, what do you get?

❑ Now try another number with a units digit of 2 (such as 22).

❑ Can you see a pattern in what happens when you halve numbers which have a units digit of 2?

❑ Write down your answers in a list:

12 → 6 22 → 11 and so on.

❑ Bring your list (and any other lists or tables you make) to school. Be ready to discuss your ideas about the patterns.

YOU MIGHT LIKE TO

Try halving numbers with other units digit values. You could make a table of your answers.

DEAR HELPER

THE POINT OF THIS ACTIVITY: is to help your child to see how doubling and halving can help with mental multiplication and division. For example, if he or she knows what 14 × 10 is, then it is easy to work out 14 × 20 by doubling. This technique will help with learning multiplication tables, since (for example) the 8× table numbers are double the 4× table numbers.

Use vocabulary such as **double, twice, 2 times, half** and **divide by 2** with your child. By the age of nine, your child should know the doubles and halves of all the whole numbers from 1 to 50 by heart.

YOU MIGHT LIKE TO: encourage your child to work out what happens when you halve a number with an odd-number units digit (such as 13). Is there a pattern?

IF YOU GET STUCK: because your child finds halving difficult, then use real objects (such as counters, dried peas or buttons) to carry out the halving or doubling physically until your child sees the patterns.

Please sign: . . . .

NAME **DATE**

DOUBLE EDGE

YOU WILL NEED: a helper, a pencil and paper, this sheet.

YOU ARE GOING TO: solve a shape puzzle using what you know about numbers.

❑ Using the drawing below as a guide, draw a side view of a cube so that three faces can be seen.

❑ Write a number inside the four edges of each face shown, so that:

1. The four numbers on each face add up to 100.
2. If there are numbers on both sides of an edge, one of these numbers must be twice the other.

❑ Look at the example below. Check that it follows the rules.

❑ Now take turns with your helper to put a number on the cube you have drawn. Remember how the cube must look when it is finished.

❑ Bring your cube drawing (and any other shapes you have made) back to school.

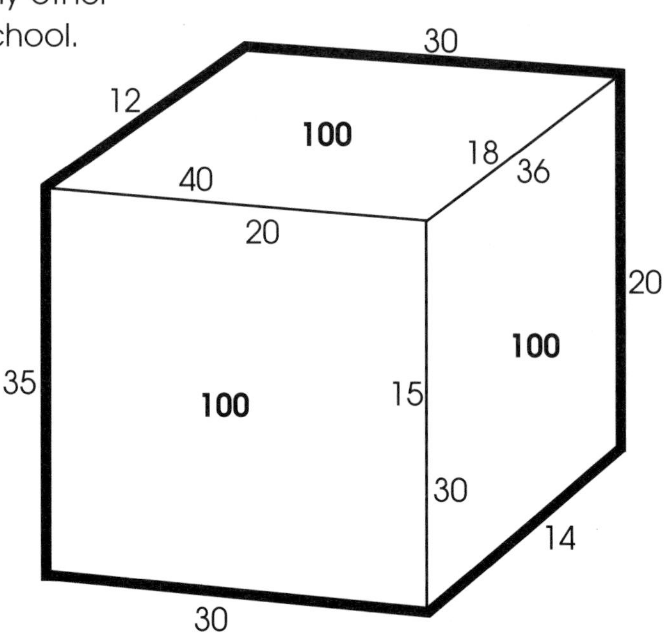

BET YOU CAN'T

❑ Make a cube where each face adds up to 1000 instead of 100.

HANDY HINT!

Look for doubles you know first, then adjust the total value for the face.

DEAR HELPER

THE POINT OF THIS ACTIVITY: is to make your child use a variety of calculation skills to solve a problem. He or she needs to plan the numbers according to two criteria: one that involves addition (or subtraction) and one that involves doubling (or halving). Your child will need to use both skills in each of his or her turns. A useful strategy for doubling a number is to look for a near number whose double you know, then adjust. For example, double 23 could be worked out as: double 20 + double 3 = 40 + 6 = 46.

YOU MIGHT LIKE TO: Work out a hexagon pattern following the two rules above.

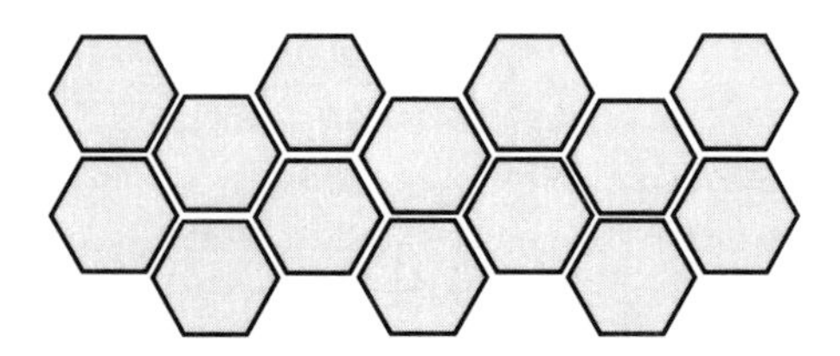

IF YOU GET STUCK: Fill in one face at a time, then place the matching edge number first on the next face. This way, you can help your child with the addition or subtraction and then focus on the doubling.

Please sign: . . . .

NAME DATE

ADDING THE TABLES

YOU WILL NEED: a helper, a pencil and paper, a timer, a knowledge of the multiplication tables from 1 × 2 to 10 × 10, a 1–100 number grid (see page 47) and a 101–200 number grid.

YOU ARE GOING TO: add up numbers from two multiplication tables.

❑ Using your knowledge of the different times tables (up to 10 × 10), what combination totals of numbers could you make? Try adding a number from one times table to a number in another times table to make one of the numbers on your two number grids.

For example, using the 2 and 3 times tables:
1 × 2 is 2, add 1 × 3 (which is 3), makes a total of 5 altogether.
Or 1 × 2 is 2, add 2 × 3 (which is 6), makes a total of 8 altogether.

❑ Ask your helper to time you to see how many totals you can make in one minute. Colour in the numbers you make on the number grids.

❑ What are the lowest and highest numbers you can make?

❑ Bring your number grids back to school, as well as your answers to the problems below.

BET YOU CAN'T

❑ Write down all the numbers between 1 and 200 that can be made in more than one way.

❑ Write down all the numbers between 1 and 200 that **can't** be made as combination totals.

DEAR HELPER

THE POINT OF THIS ACTIVITY: is to reinforce knowledge of multiplication facts and strategies for adding up two-digit numbers, and to help your child work through a problem in a systematic way.

YOU MIGHT LIKE TO: work with your child. You could each take a number from a different multiplication table, then add your numbers together. This will encourage your child to work systematically through the options. He or she will need to record the totals (on the number grids), but should not need to write down the whole of each sum.

IF YOU GET STUCK: adopt a very systematic approach. Write down the tables first, then guide your child through each addition.

Please sign: . . . .

CHANGES

YOU WILL NEED: a helper, the 'Change cards' sheet (page 39), two dice, a pencil and paper.

YOU ARE GOING TO: practise changing numbers in different ways.

❑ Cut out and shuffle the change cards. Place them face down on the table.

❑ Roll the dice and add the two digits to get the start number.

❑ Take the top change card and follow the instruction to get the finish number. Give your helper the finish number, but do not let him or her see the change card.

❑ Your helper must tell you what is written on the change card. If he or she is wrong, you keep the card and have another go with a new card. If your helper is correct, he or she keeps the card and has a go with a new card.

❑ When all the change cards have been used, count up how many cards you each have. Whoever has more cards is the winner.

❑ Be ready to explain what you did back at school. Which was the hardest change card? Why?

YOU MIGHT LIKE TO TRY

❑ Multiplying the two dice numbers instead of adding them, then playing the game as above.

❑ Making your own change cards and bringing them back to school.

DEAR HELPER

THE POINT OF THIS ACTIVITY: is to help your child to become more practised in the use of mental calculation skills such as doubling, halving, adding 9 (by adding 10 and taking away 1) and so on. These skills will help your child to carry out quite complex problem solving and investigative maths. They will build his or her confidence in using approximations to check answers.

You will notice the careful use of vocabulary on the change cards. Your child needs experience of different words that mean the same thing, so that the language does not stop him or her from carrying out the task.

YOU MIGHT LIKE TO: extend the activity by using dice with higher numbers on them (this can easily be done with sticky labels).

IF YOU GET STUCK: you may need to limit the range of cards you use. Practise carrying out one 'change' on all the numbers that you make with the dice. Introduce a new rule when your child has understood the last one.

Please sign: . . . .

NAME DATE

CHANGE CARDS

double	add 10	divide by itself
halve	take away 10	add 31
×2	plus 5	increase by 19
×3	take 5	9 more
multipy by 4	add 3	multiply by 2 and add 1
×5	subtract 1	÷1
×10	multiply by itself	minus 2
share between two groups	plus 20	

HIT THE SPIDER

YOU WILL NEED: a helper, a pencil and paper, a calculator.

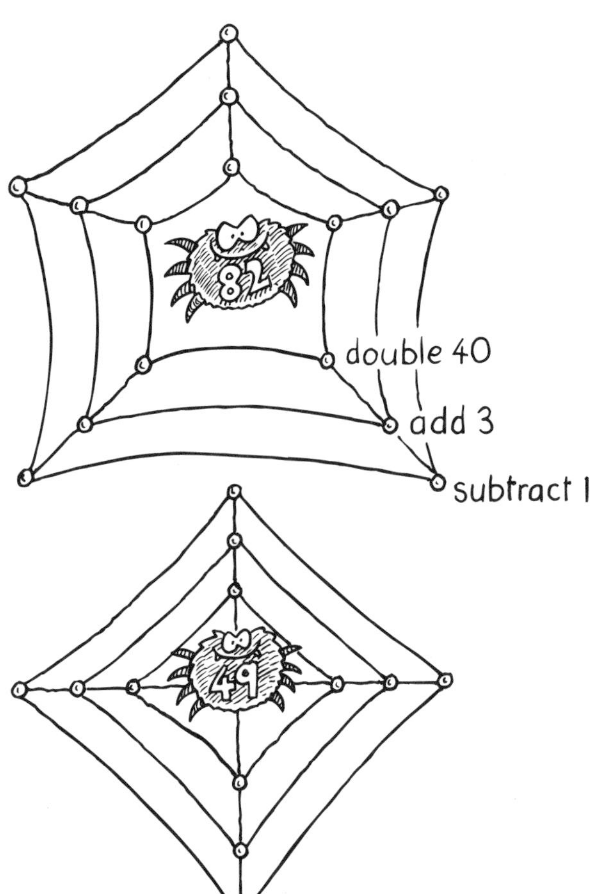

YOU ARE GOING TO: make target numbers using all the number operations you know.

- ❑ Look at these two webs. Each one has some strings leading out from the centre. Work together with your helper to complete both webs.
- ❑ To fill in the strings, you have to find different ways to make the spider number in each web. One string has been filled in for you. Each string must use at least **two** of the four number operations (adding, subtracting, multiplying and dividing).
- ❑ Check your workings, and your helper's workings, using a calculator.
- ❑ Draw webs for the spiders waiting at the bottom of the page. Work together to fill in the strings.
- ❑ Bring your answers and any targets you have made into school.

BET YOU CAN'T

- ❑ Work out the strings for the spider numbers using just multiplication and division.
- ❑ Make up some webs of your own. Try them out on your helper.

DEAR HELPER

THE POINT OF THIS ACTIVITY: is to help your child become more practised in the use of mental calculation strategies which will help him or her to carry out mathematical problems with greater ease. Encourage your child to find more than one way to make each target number. Knowledge of **number bonds** (pairs of numbers which add together to make a given total) will help with this activity.

YOU MIGHT LIKE TO: extend the activity by setting some target numbers up to 1000.

IF YOU GET STUCK: lead your child through a few possibilities to build up his or her confidence. Keep the target numbers below 100 to start with.

Please sign: . . . .

1 TO 10

YOU WILL NEED: a helper, a timer, pencil and paper, the seven cards below.

YOU ARE GOING TO: try to make all the numbers from 1 to 10 using three digits and all four number operations.

❑ Cut out these cards. Using the three numbers and any of the four operations, you and your helper (working separately) need to make as many of the numbers from 1 to 10 as you can. You can use two, or all three, of the numbers each time.
❑ Give yourselves two minutes. Now see who has made more numbers. Compare how you made the different totals.

❑ Did you make all the numbers 1–10? If not, try again. Who can make all the numbers more quickly, you or your helper?
❑ Is there more than one way to make some of the numbers?

❑ Bring your workings out, and any ideas that you have about making the numbers 1–10, back to school.

BET YOU CAN'T

❑ Try using another three single-digit numbers.
❑ Find which combination of three numbers lets you make the most numbers from 1 to 10.
❑ Work out what numbers above 10 you can make with these three numbers.

DEAR HELPER

THE POINT OF THIS ACTIVITY: is to encourage your child to develop rapid recall of number bonds and number facts in all operations of number. You should be able to make all the numbers to 10 apart from 1.

YOU MIGHT LIKE TO: try the activity as a race. This will encourage your child to use mental recall rather than trying to work everything out on paper. However, do not prevent your child from using paper as a 'board' to aid memory by jotting ideas.

IF YOU GET STUCK: encourage your child to start with the numbers 1–10 written down on paper and then write in the sums that he or she can do to make each of the numbers. When you can see a number that is giving your child trouble, ask him or her about how that number can be made: *9 take away 4 makes 5. You haven't got a 4, but (3 + 1) can make 4.* By talking through the number bonds that he or she knows, your child should work out other ways of making a number with the digits provided.

Please sign: . . . .

NAME DATE

MAIL ORDER CATALOGUE

YOU WILL NEED: a helper, a pencil and paper.

YOU ARE GOING TO: decide what to buy with your money.

❑ Look at these items from a mail order catalogue.

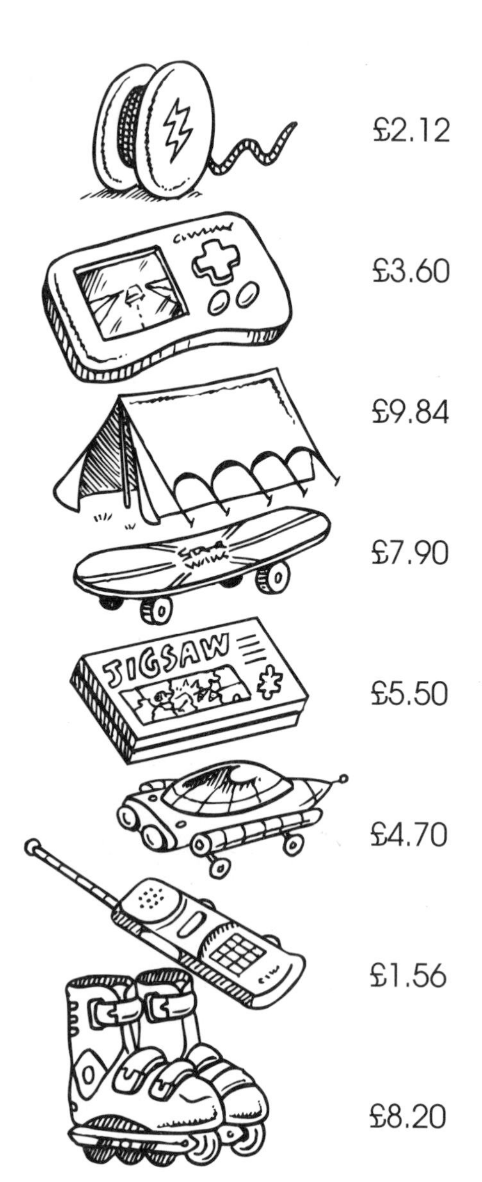

❑ You have £20.00 to spend. Work out the biggest number of items that you could buy. Try to spend as much of the money as you can. You can only buy one of each item.

❑ Make a list of the items you would buy and how much change you would have.

In January, the catalogue firm has a half-price sale.

❑ Write down what each of the items will cost now.

❑ If you bought the same items as above, how much change would you get from £20.00? Is there a quick way to work this out? Explain to your helper how you worked it out, and ask him or her to help you write down your explanation.

❑ Bring your list of purchases, and the written explanation of how you worked out the last problem, back to school.

BET YOU CAN'T

❑ Work out how many items you could afford to buy with your £20.00 if the half-price sale was still on.

❑ Work out how many items you could buy for £20.00 if all the original prices were doubled.

DEAR HELPER

THE POINT OF THIS ACTIVITY: is to help your child to solve a problem involving money, and to explain the strategies that he or she uses. Your child will need to question himself or herself about the possible strategies that he or she could use to solve the problem.

YOU MIGHT LIKE TO: help your child to organize his or her work, going through through the questions in a systematic way. Keep asking your child: *How did you do that?*

IF YOU GET STUCK:

- Ask your child to write down the prices.
- Support your child with questions, such as: *If you want to buy as many items as possible, which prices do you think you would go for?*
- Guide your child through each part of the question, so that he or she can focus on one thing at a time.

Please sign: . . . .

TARGET PRACTICE

YOU WILL NEED: a helper, a pencil and paper.

YOU ARE GOING TO: make £5.00 with small change.

❑ Look at this target.

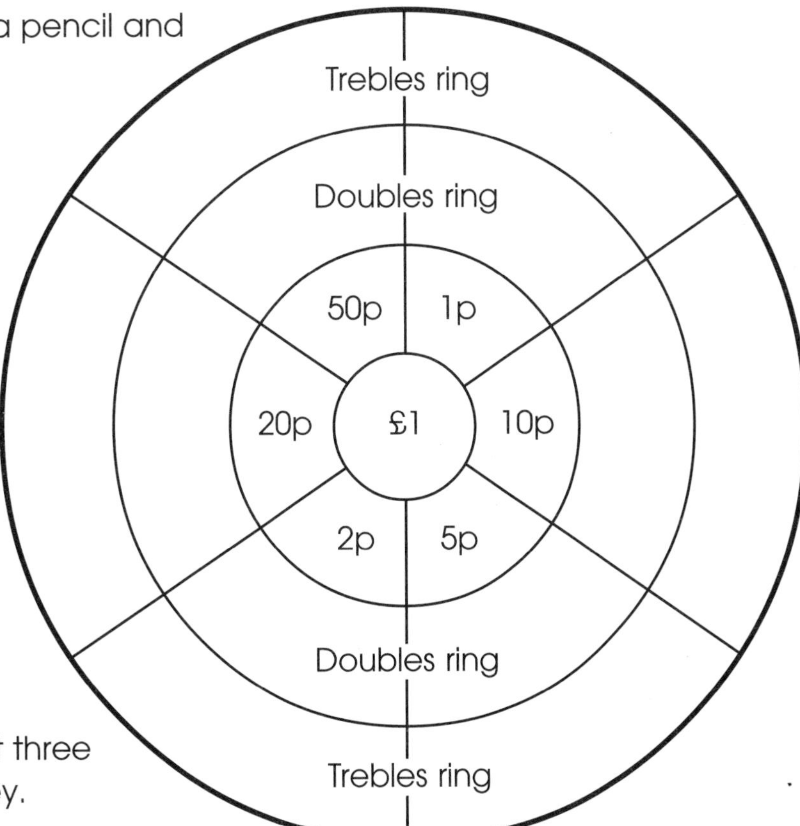

One arrow in the middle gives you £1.00. For an arrow on the board outside the middle:

- in the single ring, you get the amount of money shown for that part.
- in the double ring, you get twice that amount of money.
- in the treble ring, you get three times that amount of money.

❑ With your helper, work out how many possible ways there are of making £5.00. You can use any number of shots.

❑ What is the **least** number of arrows that you could use to make £5.00 if you could only use the treble ring?

❑ What is the **greatest** number of arrows that you might need to make £5.00 if you could only use the doubles ring?

❑ How did you know when you had found all the possible ways of making £5.00? Write down how you knew, and bring your explanation back to school.

YOU MIGHT LIKE TO TRY

Finding an easy way to record all your answers.

DEAR HELPER

THE POINT OF THIS ACTIVITY: is to help your child to develop a systematic approach to mathematical problem solving. Your child will need to be confident in his or her understanding of addition and subtraction facts, and have a basic grasp of multiplication and division.

YOU MIGHT LIKE TO: help your child to approach the problem by looking at each ring separately. Encourage him or her to work methodically through all the possible combinations.

IF YOU GET STUCK: because your child finds it difficult to do the calculations mentally, provide counters or other small objects to place on the target, marking each 'arrow'. This will make recording easier.

Please sign: . . . .

NAME DATE

THE FACTS OF NUMBER

YOU WILL NEED: a helper, a pencil and paper, a 1–100 square (see page 47), coloured pencils, counters.

YOU ARE GOING TO: make as many numbers from 1–100 as possible, using the numbers 1–10 and any number operation.

❑ With your helper, choose three numbers between 1 and 10. Then list all the numbers you could make by combining them in pairs with any one of the four operations: +, –, × or ÷.

For example, with 5 and 2 and 6:

$5 + 2 = 7$	$6 + 5 = 11$	$6 + 2 = 8$
$5 - 2 = 3$	$6 - 5 = 1$	$6 - 2 = 4$
$5 \times 2 = 10$	$6 \times 5 = 30$	$6 \times 2 = 12$
$5 \div 2 = 2\frac{1}{2}$	$6 \div 5 = \frac{12}{10}$	$6 \div 2 = 3$

❑ Check each other's work. Whoever has more answers gets a counter. If you both have the same number of answers, you both get a counter.

❑ Now shade all the numbers that you have made on the 1–100 square. Are there some numbers that are not shaded? Make a list of these.

❑ Keep trying different combinations of three new numbers. Mark off all the numbers that you can make.

❑ Bring your shaded number square and the list of numbers **not** made back to school.

BET YOU CAN'T

Work out how many numbers on a 1–100 square can be covered using **all** the combinations of three numbers from 1 to 10.

DEAR HELPER

THE POINT OF THIS ACTIVITY: is to encourage your child to work in a logical way through an investigation. It is important to be methodical when combining each number with each of the other numbers.

YOU MIGHT LIKE TO: encourage your child to use larger starting numbers and a 1-400 or 1-900 number square.

IF YOU GET STUCK:

- Help your child to be organized by first working out the various combinations that are possible, then letting him or her find the answers.
- Approach the activity in a slightly different way. Do all the additions first, then all the subtractions, and so on.

Please sign: . . . .

UNDER THE COVERS

YOU WILL NEED: a helper, some small paper rectangles, the grid on this sheet, pencils.

YOU ARE GOING TO: make up number sentences for a puzzle.

- ❑ Write a different number in each of the rectangles in the grid below. Then cover each number with a paper rectangle.
- ❑ Write on each paper rectangle a way of making the number that is written underneath. This might be a number sentence using any of the four operations: +, −, × or ÷.
  For example, for **34**, you might write **39 – 5**.
- ❑ Now challenge your helper to work out the number underneath each cover.

- ❑ When your helper has guessed all the numbers correctly, shuffle the covers and then see if you can put them all back.
- ❑ Bring your number grid and covers back to school to try on other children.

125				
	34			16
		129		
			12	

YOU MIGHT LIKE TO TRY

Letting your helper fill a grid with numbers and place covers for you to work out.

DEAR HELPER

THE POINT OF THIS ACTIVITY: is to help your child become more practised in the use of number operations. He or she will need to read and write number sentences, using the symbols associated with the four operations.

YOU MIGHT LIKE TO: try out some examples of ways to make different numbers before you start. Encourage your child to be inventive with the number sentences.

IF YOU GET STUCK: try limiting the number sentences to one operation (initially addition) and keeping the numbers small. You can then try subtraction, multiplication and division using fresh grids.

Please sign: . . . .

NAME DATE

WHAT'S YOUR RULE?

YOU WILL NEED: a helper, pencil and paper.

YOU ARE GOING TO: make up rules and apply them to numbers.
Jill's rule for changing two-digit numbers is: *double the units digit, then add the tens digit to it.* For example:

Start number	Rule applied	Finish number
13	Double the 3 units makes 6, add the 1 ten	7
15	Double the 5 units makes 10, add the 1 ten	11
21	Double the 1 unit makes 2, add the 2 tens	4

❑ Think of some more rules for changing numbers. You can use any operations: +, −, ×, ÷, halving, doubling and so on.
❑ Give your helper several start numbers and finish numbers. See whether he or she can work out your rule. Try your other rules.

❑ Bring your rules, start numbers and finish numbers to school to try out on other children.

BET YOU CAN'T
Find a number that doesn't change when Jill's rule is applied.

DEAR HELPER

THE POINT OF THIS ACTIVITY: is to encourage your child to use, read and write different number operations. Ask your child to show you how Jill's rule works by writing down some new number sentences with appropriate symbols.

You may need to give your child some time to work out his or her own rules; after playing, ask for a few further examples with each rule applied to check that he or she is using the operations confidently.

YOU MIGHT LIKE TO:
- Give your child some different start numbers and ask him or her what the finish number would be. You can use Jill's rule or one of your child's own rules for this.
- Give your child some rules to work out (keep them simple to begin with).

IF YOU GET STUCK: because your child is having difficulty thinking up rules, give him or her some simple rules to apply to a sequence of numbers so that he or she can practise using the different operations. If another helper is available, your child could work with him or her to set rules and provide you with start and finish numbers.

Please sign: . . . .

NAME DATE

DIGIT CARDS AND NUMBER GRID

1	2	3	4	5
6	7	8	9	0

1	2	3	4	5	6	7	8	9	10
11	12	13	14	15	16	17	18	19	20
21	22	23	24	25	26	27	28	29	30
31	32	33	34	35	36	37	38	39	40
41	42	43	44	45	46	47	48	49	50
51	52	53	54	55	56	57	58	59	60
61	62	63	64	65	66	67	68	69	70
71	72	73	74	75	76	77	78	79	80
81	82	83	84	85	86	87	88	89	90
91	92	93	94	95	96	97	98	99	100

Dear Parent

We all know that parents are a crucial factor in their children's learning. You can make a huge difference to your child's education. We are planning to send home some activities that fit in with the maths we are doing in school. The activities are designed for your child to do with you, or another available adult. You do not need to know a lot of maths in order to help your child.

These are not traditional homework activities. It is important that your child first explains the activity to you. Each activity will have been explained thoroughly in school. Then do the activity together. By sharing these activities with your child, you will be helping to develop her or his mental maths. And as a result of being given that all-important attention, your child is more likely to become confident and skilled in maths.

We hope, too, that these activities will be fun to do – it matters that children develop positive attitudes to maths. If you are particularly nervous about maths, try not to make your child nervous too! If your child is having difficulties, look at the 'If you are stuck' suggestions which are provided on each activity sheet.

After completing each activity, your child will usually have something to bring back to school. However, sometimes there may not be anything written down to bring back – your child is doing mental maths, so all the work may be in your heads!

If you have any problems with or further questions about any of the activities – or about any of the maths being covered – please do let us know at school. We do very much value your support.

Yours sincerely